Experts versus Laymen

Experts versus Laymen

A Study of the Patsy and the Subcontractor

Barney G. Glaser

Routledge
Taylor & Francis Group
LONDON AND NEW YORK

Originally published in 1972 by The Sociology Press.

First published 2015 by Transaction Publishers

2 Park Square, Milton Park, Abingdon, Oxfordshire OX14 4RN
605 Third Avenue, New York, NY 10017

Routledge is an imprint of the Taylor & Francis Group, an informa business

First issued in paperback 2020

Library of Congress Catalog Number: 2015023369

Library of Congress Cataloging-in-Publication Data

Glaser, Barney G.
Experts versus laymen : a study of the patsy and the subcontractor / Barney G. Glaser.
pages cm
Includes bibliographical references.
ISBN 978-1-4128-5717-8
1. House construction. 2. Construction contracts. I. Title.
TH4815.5.G55 2015
692'.8--dc23

2015023369

ISBN 13: 978-1-4128-5717-8 (pbk)
ISBN 13: 978-0-87855-217-7 (hbk)

To Anselm

I wish to express my appreciation for the editorial work of Anselm L. Strauss and Kelley Weisberg and the typing of Pat Rodden and Elaine McLaren.

Contents

Preface

This monograph is one additional and related volume in the body of work being produced by Barney G. Glaser and Anselm L. Strauss. The books now number twelve.* It is important to locate EXPERTS VERSUS LAYMEN within the sequence of these books, to show how it fits and to what problems it is addressed. I shall try to locate it regarding the two main tasks of our work: (I) the rendering of society in terms of social interaction, social structure and social process and (2) the development of grounded-theory methodology and style. Lastly in (3) I indicate the wide potential use of this book.

*1. Strauss, Anselm, MIRRORS AND MASKS (Glencoe. Ill.: Free Press of Glencoe, 1957) and reprinted, 1970, The Sociology Press, San Francisco.
2. Strauss, Anselm, IMAGES OF THE AMERICAN CITY (New York: Free Press 1961)
3. Strauss, Anselm, Leonard Schatzman, Rue Bucher, Danuta Ehrlich and Melvin Sabshin, PSYCHIATRIC IDEOLOGIES AND INSTITUTIONS (New York: The Free Press, 1964)
4. Glaser, Barney, ORGANIZATIONAL SCIENTISTS: THEIR PROFESSIONAL CAREERS (Indianapolis: Bobbs–Merrill, 1964).
5. Glaser, Barney, and Anselm Strauss, AWARENESS OF DYING (Chicago: Aldine Publishing Co., 1965)
6. Glaser, Barney and Anselm Strauss. TIME FOR DYING (Chicago: Aldine Publishing Co., 1967)
7. Glaser, Barney, ORGANIZATIONAL CAREERS (Chicago: Aldine Publishing Co., 1966).
8. Strauss, Anselm, THE AMERICAN CITY (Chicago: Aldine Publishing Co., 1968)
9. Strauss, Anselm, THE CONTEXTS OF SOCIAL MOBILITY (Chicago: Aldine Publishing Co., 1971)
10. Glaser. Barney and Anselm Strauss, STATUS PASSAGE (Chicago: Aldine Publishing Co., 1971)
11. Strauss, Anselm and Barney Glaser, ANGUISH (San Francisco: Sociology Press, 1970)

12. Glaser, Barney and David Crabtree SECOND DEEDS OF TRUST (Mill Valley, California: Balboa Press, 1969)

(1) Rendering Data: The task in this particular book was to write about interaction, to develop a style that would densely render interaction and to show how interaction varied under specified structural conditions. For the PATSY, social process is not paramount, although interaction is always treated processually. Thus it falls between AWARENESS OF DYING and TIME FOR DYING in design and accomplishment. AWARENESS was our first attempt to write about interaction under carefully specified structural conditions (awareness contexts, etc.). TIME FOR DYING was more sophisticated in that we had come to write about interactional materials as they are related to ongoing structural process—social structures n process. The PATSY book was planned and data collected right after AWARENESS went to press. TIME FOR DYING was planned earlier but written two years later. PATSY was written in 1970-71 but remained faithful to its 1965 planning, as this is what the data provided. Being a 1971 product, it was done with much greater methodological skill than TIME FOR DYING. One of PATSY'S principal features is the *dense* treatment of interaction and of the complex interplay of strategies and counter-strategies. Its degree of density is partly made possible by the more restricted types and number of actors as compared with those who appear in TIME FOR DYING.

Predating AWARENESS were three books. The first, MIRRORS AND MASKS was designed to stimulate attempts at the grounding of theory which would conjointly handle social structure, process and interaction; it also set the general style of the later publications. In IMAGES OF THE AMERICAN CITY, a process (American urbanization) was at the center of focus, with analysis done of some of the conditions (region, era and type of city) of this process. (Years later, a related type of analysis of a process—the professionalization of an occupation—was published in a long paper "The Structure and Ideology of Nursing," with an emphasis on various structural conditions, some persisting and some changing over time, which have affected the basic process studied.) In PSYCHIATRIC INSTITUTIONS AND IDEOLOGIES, the main focus was on social process, with some effects on interaction and on the maintenance of social order. Specifically that book dealt with the careers (a type of social process) of health professionals and patients as they intersected—as a result, social order was continually being renegotiated (a type of process). Thus AWARENESS had moved the style forward by developing a full-scale analysis of interaction under specific structural conditions, and by the development of theoretical sampling as well as the type of comparative analysis associated with that sampling. Up to this point, social process had been finely developed principally along the lines of "careers" PROFESSIONS, WORK AND CAREERS reprints several papers and chapters of books which deal with these topics and with social processes. Glaser had also contributed to this development with Strauss' counsel by publishing ORGANIZATIONAL SCIENTISTS: THEIR PROFESSIONAL CAREERS before AWARENESS was conceived (although after the terminal care project was underway). We returned to social process again in TIME FOR DYING by developing structural process as well as careers (trajectories).

(2) Grounded-Theory Methodology: After TIME FOR DYING, we set several tasks for our work, broadening the above renderings through the further development of methods for generating grounded theory. We developed –by generating substantive theories- the various renditions of social process, interaction and structural process and conditions. The above noted books—and the PATSY—present substantive theory—theory about a specific area. (IMAGES OF AMERICAN CITIES also had had the task of developing substantive-theory categories by which others might, if they wished, generate grounded substantive theories of urbanization.) After AWARENESS, we published a methodology book dealing with the generation from data of substantive theory—THE DISCOVERY OF GROUNDED THEORY. In it, we also set ourselves directions for generating other forms of substantive theory and for generating and presenting formal theory—theory about a conceptual rather than substantive area. We also mentioned the application of grounded theory.

Thus, in ORGANIZATIONAL CAREERS: A SOURCEBOOK FOR THEORY, formal theory was generated from a set of substantive theories done on the social process level. In THE CONTEXTS OF SOCIAL MOBILITY: IDEOLOGY AND THEORY, a substantive theory on mobility was generated from data and by using the formal theory as presented in STATUS PASSAGE (which was written but not yet published). Another aim of the mobility book was to transcend the limitations of current theories of mobility—and avoid their deep rhetorical biases—by theoretical sampling and comparative analysis and using historical data, as outlined in DISCOVERY. The task in STATUS PASSAGE had been to develop a pure formal theory, both from other grounded theories and from data pertaining to a fundamental social process. In this book, we also *linked* a formal theory of status passage with that of awareness contexts.

On the applied level. ANGUISH: A CASE HISTORY OF A DYING TRAJECTORY had the task of showing how grounded substantive theory could be used to provide theoretical explanations for one case history, which itself fed back to inform the substantive theories developed in the two previous dying books. We also used a partly developed, grounded theory of pain management. For another book, SECOND DEEDS OF TRUST: HOW TO MAKE MONEY SAFELY (Glaser and Crabtree), an effort was made to generate a substantive theory of investing, tailored to be readily applicable to managing that kind of investment.

In the relatively near future, we anticipate publication of a second methodology book: THEORETICAL SENSITIVITY. In this we shall outline the operations of generating theory that we have developed in our writings and in seminars since the publication of DISCOVERY. The current book, THE PATSY, has obviously benefited from this advance in our methods. Our students and colleagues have helped us very much by posing many questions, problems and critiques of grounded theory methods. Following THEORETICAL SENSITIVITY, we plan a volume on Chronic Illness and another on Negotiation. The former wilt develop a broadly based substantive theory, which is not yet formal theory, of high applicability to the relevant substantive area; the emphasis being on multiple processes. The monograph on negotiation will present a formal theory of negotiation and give its application to a series of substantive areas via the analysis of case studies.

(3) Dimensionalizing: Before closing this preface it is important to emphasize the dimensionalizing aspect of this book. By this I mean the ability to take the categories, properties and propositions generated here to any area of expert-layman relationship or servicing relationship and use them to develop emergent explanations of what is going on. In doing this the sociologist adds, subtracts, and modifies the dimensions to suit his specific substantive context. Thus, the forthcoming theory has tremendous possibilities for guiding substantive research into professional-client and nonprofessional–client services. It also has potential for application in controlling these relations. Lastly, it can easily be raised to the level of formal theory of expert-layman relations by comparatively analyzing these several areas, as I attempt to indicate in the last chapter.

Needless to say a book on this area will be useful for handling subcontracted repairs or improvements, which is far more usual than building one's own home.

I

Playing Patsy

We live in a society of laymen who devotedly bow to the wisdoms and whims of experts by complying with their orders and following their advice. The patsies (laymen)—the know-nothings—turn to the experts—the know-it-alls—for help, service and work. The patsies have no training and little experience and the experts have both. The patsies have no time, and usually no inclination to find it, to gain the training and experience necessary for handling their problems. Accordingly they turn to the experts who make a business and sometimes a profession of putting the requisite time into those kinds of problems and who can assure a timely resolution and a good measure of quality. In turning to the expert, the layman usually gives what the latter claims, that is, almost unconditional compliance with his conditions for the timing, costs, and quality of work which will bring about a satisfactory resolution despite any by-product stresses. The patsy will give, in short, the expert power over their working relationship in anticipation of a successful outcome to his problem. The expert, then, clearly has in his favor the balance of *power symmetry* over their continuing relationship. The expert can direct the work to be done during the period in which he holds his power.

To be sure, the balance of power symmetry and its constitutive sources and consequences vary for the different expert-layman relationships, and vary over the course of their relationship. We, in sociology, are familiar with studies of the professional-client relationship (especially for doctors, nurses, social workers, teachers and lawyers) in which the power initially is asymmetrically in favor of the professional and continues so during most of the course of the relationship. Thus, a doctor or lawyer generally has a great deal of power—given and acquiesced to by the patient or client—for directing medical recovery or a legal matter. Patients or clients who do not accept the asymmetrical relationship offered by their doctor or lawyer make interesting stories, but are still infrequent enough not to disturb the general pattern.

This book presents some theory on the expert-layman relationship, based on one patsy's dealings with approximately 200 subcontractors at various stages of home construction. In contrast to the above mentioned professional-client relationships, the balance of power is relatively symmetrical in the patsy-subcontractor relationship. Also, the balance subtly shifts back and forth between expert and patsy. The latter can pick up a slight edge over any given subcontractor through his ability to comparatively bid proposals for work—on the dimensions of time, quality and cost demanding that high quality jobs will be done on time during the course of the job, and can withhold funds until such work is satisfactorily completed. Often, the patsy will relinquish or lose this edge of power either because he is unaware or is outmaneuvered by the subcontractor. Their relationship is a *continual contest* over who will have the edge of power over the entire course of work. The contest does not occur when the patsy relinquishes his chances for control, which many do; the subcontractor, of course, *never* relinquishes his.

The Core Categories

The theory to be presented will emphasize the social processes central to getting a private home built. (These processes also occur in the remodeling or repairing of any aspect of a house or of an apartment.) The theory will lean heavily on several interrelated core categories which have been discovered to be most relevant in managing the working relationship between patsies and subcontractors. One is the *power symmetry,* this by inference, very relevant to understanding any expert-layman relationship. A second core category is the *competitive bidding system.* I shall explore the nature of this system which allows the layman to bid experts against each other so that he may choose the best qualified one for his own control and protection in the ensuing relationship. Also, I shall consider in Chapter IX the conditions under which different types of expert-layman relationships either foster comparative bidding or make it considered as illegitimate as with the bidding of doctors or lawyers. Power symmetry is, of course, one such major condition. The more power that experts generally have, the less they allow legitimate comparative bidding for their services.

A third core category is *elsewhereism,* which pertains to the multifaceted effects upon the expert's continuing work relationship with the patsy which flow from other laymen's and expert's demands for his time and services. Thus, the patsy is in competition with people elsewhere; and he has little or nothing to go on with respect to how to compete in order to get the expert's services on time and quality at a reasonable cost. Practically the only time the layman will hear details about these competitors is when the expert refers to them during his use of an excuse tactic for delaying the patsy so that he will await his turn for services or work. Many of us are familiar with this source of delay when we try to see physicians. In particular, we shall be interested in how a patsy must pay for a viable place within this system of competition with people elsewhere.

A fourth core category is the patsy's *generalling* of the total job. With respect to his subcontractors, this refers to: (1) comparatively bidding the subcontractors' offers, (2) administering the temporal ordering of the job which entails supervising each subcontractor over time and articulating many jobs at the same time and in sequence, (3) inspecting, evaluating, trouble-shooting, and hence controlling the quality of work—then paying for it (bookkeeping). Concerning generalling, emphasis therefore will be put on the categories of *cost control, quality control* and the *temporal control* of jobs as the power symmetry shifts back and forth—particularly when the patsy's edge is continually undermined by *elsewhereism.* (Generalling also refers to many other necessary facets of the total job, which will be outlined later in this chapter, but which will not directly concern us in this book.) In the typical expert-layman relationship, the expert will general the work to be done; here, however, we have a layman generalling expert subcontractors, who in turn run their crews of unknown.

The patsy who has authored this monograph has been involved in the subcontracting necessary for building three homes, on two of which he did the rough carpentry. For each house, there were about 72 subcontractors involved in bidding and 24 involved for the actual jobs. The author also has interviewed numerous other patsies about their own experiences with subcontractors. All were instantaneously ready to share them, as they were very enjoyable to relate. Thus, this study is based mainly on participant observation which yielded the data I shall render into grounded theory, according to the qualitative analytic processes set forth in. *The Discovery of Grounded Theory.*

Now let us consider why a patsy should bother generalling his own job, and then how he gets started.

Why Bother?

The notion of turning oneself, a layman, into playing patsy with a number of subcontractors, usually raises the question: Why bother? Why not just turn the building or remodeling over to a general contractor and let him handle the subs? Isn't it worth paying something extra to let an expert handle these experts so that the patsy is not bothered? Afterward the layman can just start using what has been built for him. The answer is, of course, *yes indeed,* let the general contractor handle it but . . . make sure that he is "good" or in the end you (the layman) may be *more* bothered during the job than you expected. And, if you are going to be bothered anyway there are several structural reasons why the patsy might run the show himself.

The layman can become quite upset by the additional costs, sometimes prohibitive, of using a general contractor—so he can cut costs by not using him. The same calculation applies to the quality and timing of work: the chances are they can be made more favorable by skipping the general contractor. For these structural advantages of control, the patsy must put in time, (which, of course, is often a structural disadvantage given his own work situation). The size of some repair and remodeling jobs is just too small to require a general contractor; then dealing directly with the subcontractor makes sense. Lastly, playing patsy is fun and brings satisfaction to many a layman; so why should he pay a general contractor and miss this joy? Let us consider, in turn, the manifestations of each of these sources for a patsy generalling his own jobs.

Cost Control: Cost control refers to how much a job costs, how the money is distributed throughout the diverse costs of the job, whether the price is good, and to being able to hold back payments to insure satisfactory work. The general contractor, at minimum, puts 10% overhead and 10% profit (21%) on top of subcontracting bids. Thus, in bypassing the general, the patsy can immediately save 21% on costs, and for many a limited budget this makes the difference in affording to remodel or build a home.

Even if the patsy is not forced financially to doing the generalling, he still may find doing it quite beneficial in terms of how his budget will be distributed throughout the construction. Money goes in three directions: contractors, labor and materials. The less that goes to contractors, the more can be put into labor and materials—and this most laymen quickly realize, is where the money counts most in the total job. Thus, bypassing the general contractor (even subs sometimes if the patsy knows a competent workman) can allow for better materials and sufficient money that will allow labor to take the time to do quality work for less over all costs.

In a recent issue of *Time* magazine, an article underscored the resentment that people feel at the high costs of generals and seeing the money that does not reach the actual job.

> The man who resents paying the high price of home repairs has few alternatives. Some save by acting as their own contractors. buying materials at the contractors' discount and employing moonlighting carpenters and electricians. The moonlighters generally charge only their actual wage rate, plus perhaps a dollar an hour.

This quotation also bears on the problem of adequate bidding to obtain the best prices available and to avoid the exploitation that causes resentment. Many general contractors do not bid subs as well as they might to achieve good prices. There are several reasons why they do not. A general contractor may wish to save the time, or he may have some favored subs that he works well with, and uses frequently and has referral exchanges with. Chances are, then, that he is not getting quite the best prices for the job. Laymen who can pay his total bid, and do not know about the bidding system, may never know why, or if, his prices are high for this reason. They may guess and worry a little bit, but they certainly do not contact subcontractors directly. However, this is not really exploitation or "gouging" by general contractors. Simply, it is not using the system to its fullest potential for keeping costs down. The patsy who acts as his own general can get the most cost control out of the bidding system.

Gouging refers to a very high level of costs which are far out of line with those of other general contractors, and it usually results in shoddy work. Gouging by general contractors is more a myth than a reality. If laymen comparatively bid general contractors, they quickly catch a gouger. But they may find it very hard to catch the general whose price is not based on adequate bidding of subcontractors: many generals, as we have said, do not bother; and total cost variations among the reputable generals tell little when variations in quality and materials must be taken into consideration. Needless to say, some generals do go over their subcontractor's bids carefully with their prospective client. In this way, the patsy can judge how the money will be spent, and that the general is not overcomplicating or overdoing simple jobs and thereby raising costs.

The ability to pay directly and especially to withhold final payments, provides a patsy with the assurance that each subcontracting job will be done with expected quality and will be finished properly and on time. This kind of financial control is invaluable because once a sub is paid in full, getting him back to correct quality or mistakes or to finish can be quite difficult. When this power is in the hands of a general contractor, subs may be let off the job too easily, simply as part of the general's bookkeeping practices or because his attention is drawn elsewhere to other jobs and people. And it is even harder to get the sub back when the general cannot pay him if the latter runs out of money. When this happens, and it happens all too readily, the layman is stuck with the agonizing task of having to hound the general to hound the sub. But his own power is clearly minimal so he must wait while living with unfinished or uncorrected work. He cannot even hold out funds on the general as this would only paralyze the payment of all subcontractors on the job construction rather than punish the one who is derelict in his duty. Also, usually the lending company that puts up the construction money will pay the general on a regular schedule, no matter what "little" bugs still remain. In short, when the layman plays patsy to subcontractors rather than to a general contractor, then the costs to which he commits himself do not have to be met until the job which he bargained for is satisfactorily completed.

Quality Control: The essential factor in quality control is closeness to the job: the closer the more control. In bypassing the general contractor, the patsy, by getting closer to the actual work, can exert more control. This, of course, is of great concern since he will live with the quality for many years to come. The ceding of this control to a general may, for structural reasons, jeopardize quality.

During the subcontracting, bids for prices represent negotiated quality. To make the total bid (a composite of many subcontracts) more reasonable, competitive or palatable, a general contractor may negotiate somewhat less quality for a lower price, while the layman is none the wiser. Also, quality means providing the necessary time for the proper effort; since time is labor, hence money. Rush jobs are not good jobs as a rule, and since rush jobs are required for lean prices, the layman can easily become dissatisfied when he discovers his error but too late for rectification. Thus, in the bargain the layman (by using a general) can easily yield sufficient *temporal control* over jobs so necessary for securing quality. If the sub does not finish in time, he may insufferably (for the layman) delay completing it, since further work means working for nothing, as the sub has overrun his calculation for the bid. Also, the chances are that the general will pay him anyway for work already done. The new or remodeled homes where particular jobs have been left indefinitely uncompleted is legion! The layman perpetually calls a general who does *not,* in turn, call his sub though he avows that he will. The patsy does not know whether the general is really drawn elsewhere to more lucrative jobs, or through poor bidding and poor running of the job has no money left to use for finishing odds and ends of the construction.

In generalling the job himself, the patsy can more closely control these financial and temporal aspects directed at securing the quality, by paying somewhat more for the time and material it requires while actually still preserving part of his dollar savings by not using a general. This statement does not mean that many generals do not maintain quality control for their clients. It does mean, however, that a layman can turn to generalling his own job to ensure quality and insure against the possibility of obtaining a general who is lax in this regard. Discovery of this laxity is invariably too late, even after reviewing other work of the general—conditions change for each job.

During the job, quality is controlled by continual inspection, evaluation, and above all, acceptance of completion which provides the sub with closure. To run the job—achieving the quality bargained for and receiving it on time—clearly involves the world of strategy and counter-strategy in a context of elsewhereism. Being close to quality control means that the patsy participates directly in the work with the subcontractor. The patsy can be sure jobs are done right *the first time,* by preventing errors and making sure that designs are adhered to. He can make all the phone calls to the subs and the discrete multitudinous daily decisions that are so necessary to keep the job going right. He can withhold payment and psychological closure until the jobs are done right—even when correcting inadvertent or sloppy errors—at the subcontractor's expense.

Some laymen say that this is the aspect of the construction they wish to avoid by hiring a general contractor. Some find that when things go wrong or questions arise that they are forced into this world of relationships anyway. Yet, then they have not the power they need to be effective, being organizationally one level away from being able legitimately to give orders to the subcontractors. They can scream, but the sub waits to hear his orders from the general who is in charge of work and pay.

The layman, then, calculates his approach to the job accordingly. Should he turn patsy and control quality himself, or which are the available general contractors he can trust to ensure quality? And will he be forced anyway to put in much of his time that he is supposed to be saving to supervise quality; if so, why should he put it in with diluted power? The calculation of time is crucial. Playing patsy takes time and sometimes, with home construction, it takes full time in the beginning stages; but time is valuable.

Thus, the patsy must calculate how much he saves and loses to take time off from his regular job to build a house. Given the costs of construction today, he has to have a very good job actually to lose more than he gains in saving on construction costs. If he only subcontracts the house, and does no actual building himself, like the general, he need not be there all the time, although a daily review is often mandatory during working hours. If his own occupation can stand this kind of part-time disappearance for four to six months (as many can), he will have no great problem. He also has weekends and nights for bidding, talks with subs, and bookkeeping. And, he has vacations for the intensive beginning of getting a house off the "ground."

Even if his own job cannot stand this part-time disappearance or taking off six months to completely build a home, it still may be worth his while to take the time anyway to ensure quality and cost control. After all, he is putting the time into the home he will live in! If his own home was not that important, chances are that he would not wish to have it built to begin with. Otherwise, he would shop for a '"spec" or older house. Also, when the nature of his regular job permits, putting his own time into generalling the house may allow him to be more flexible about the pace construction than general contractors can allow themselves. And, temporal flexibility is the major source of pacing that ensure quality. Repairing and remodeling present less temporal demands and, therefore, are less problematic for dispensing with the services of the general contractor.

Fun and Satisfaction: A vital aspect of the calculation about how much time a patsy can put into building his own home is whether or not he wishes the fun and satisfaction to be gained thereby. Many patsies in building or remodeling their *own* home, have to be involved as completely as possible every possible minute. It's to be their home—and they want it exactly their way, and wish to grow it and to grow with it. It is the central joy of their life—their own home and its construction. On this score alone they will take the time to play patsy with subcontractors and will probably do much of the work themselves, particularly the rough carpentry. Because of finances and regular jobs, some will build only on weekends. Others will quit work or take a leave of absence. Whatever the conditions for finding the time, they—like the present author—will not pay a general contractor to have their fun for them; they find the time.

Size of Job: Securing home repairs is something that every family faces. The most typical repairs are in plumbing, electrical, painting, cabinets, roof and heating. Securing these repairs oneself is even necessary when renting an apartment. To extrapolate from *Time* magazine, calling repairmen is on the increase.

> As housing prices climb and mortgage loans become costlier and scarcer, more and more people find themselves forced to stay in older houses for longer than they would like. Sooner, rather than later, pipes crack, paint peels . . . -and homeowners have to face up to the often traumatic experience of calling that new aristocrat of the U.S. labor force, the repairman.

The size of these jobs often do not warrant a general contractor. The decision is usually whether to do-it-yourself or call in a subcontractor. Do-it-yourself electrical, plumbing and roofing are usually too difficult for the average layman, so he must subcontract. The thrust this book deals with is how the patsy subcontracts so as to avoid or at least minimize what *Time* refers to as "maddening delays, shoddy workmanship, and startling expense" in characterizing home repair.

Generalling

There is much that goes into generalling a construction, and in reduced degree, a remodel or repair job, that will not be in focus in this book except tangentially to our discussion of the patsy-subcontractor working relationship. Therefore, in this section will be briefly considered what we shall then assume the patsy has taken care of in order to gel the job off the ground and to muster the resources to keep it moving.

First of all, the patsy needs a piece of land that is a ready building site, which initially many lands are not. This means he must have water, power, gas (usually) and electricity and sewerage facilities. Many areas accept septic tanks when sewerage lines are not available. The particular lot, however, must pass a health department inspection on whether the land can hold a proper septic facility for the size of the house. Thus, to be a building site the land must have a septic tank permit. In some areas where neither gas nor electricity can be obtained, butane gas may be acceptable. Water lines may have to be put in, or old lines brought up to standard. Power poles may have to be installed to bring electricity up to the lot. A survey should be done to show all concerned where the lot is and how the building is located on it. Quite often a survey shows that the lot is different than imagined to be.

Once a lot is considered a building site, then the patsy must secure a design for the house he wishes, which will fit on the lot and for which he has had plans drawn. Architects, of course, do the whole job. However, if the patsy designs his own house he can then take the general plan to an architect, general contractor, or engineer to have them formulate the structure and draw specific plans. Or, if the patsy knows building codes and the standard framing structure, he can take his design to a draftsman who will draw the plans. Lastly, the patsy with sufficient talent can do all these steps himself. In replacing these experts, his talent need not equal theirs by any means. He need only know what he wants and know enough about codes and construction to draw enough of a plan (poor as it may be) to obtain a building permit and financing. The patsy does not have to please a client, as an expert does.

The patsy does have to produce a plan with sufficient "know-how" embedded in it to convince a building department and a lending institution that he has a feasible project for a bona fide building site. Both lenders and building inspectors will make on-site inspections with plan in hand for this judgment.

With his plans, the patsy applies at the local building department for an "owner-builder" permit. Upon receiving this permit he, in effect, legally becomes his own general contractor and may legitimately begin to get the job going. He needs many basic items at first. The health department will require a toilet facility on the property. There is always at least one local company which rents self-contained chemical "outhouses" for construction. These little houses can also conveniently double as a tool shack for the minimum amount of tools left on the job. Usually all workmen carry their tools home with them. Tool shacks can also be rented or easily built. In addition the "outhouse" can be used as a phone booth until the construction is advanced enough to switch the phone to a place in the house that can be locked at night. The phone is an absolute necessity for subcontracting and for ordering materials during the day's work. The triple use of the "outhouse" is one of the patsy's first steps toward cost control. Also, the patsy may rent a power-pole to which it is legal for the electric company to hook electricity. He could build the pole, but renting one that already meets code specifications is cheaper and faster. In mustering these resources the patsy need only start "calling around" and by the nature of his queries will be "'passed around" to the companies that are engaged in renting and supplying these facilities. This referral system works well, and delivery is the "next day." The few companies engaged in providing these supplies understand the immediacy of need for them. They have a "like yesterday" attitude, so the patsy is not delayed by them.

While he is obtaining these basics the patsy is also applying for financing from local lenders, and visiting material suppliers to see if he can arrange discounts and where the best lumber can be found, obtaining the necessary insurance (liability, fire and workman's compensation), and rounding up the tools that he will use. Lenders, material and tool suppliers, and insurance can, like subcontractors, be bid comparatively for cost and quality control. In accomplishing these tasks, the patsy must keep in mind the degree to which he plans on generalling his construction; that is, how will he locate himself in the natural division of labor of home construction. This location partially dictates his needs for starting the job and keeping it moving.

Now it is important to note briefly, but in somewhat more detail, the patsy's problems in obtaining the building permit and make clear his condition in possessing the requisite temperament resources to subcontract adequately and locating himself as general on the job. Then we shall discuss when is the favorable season of the year in which to build.

The Building Permit: As a patsy, I chose to have a designer draw the house and draft the plans in as brief a manner as possible. The intent was to play patsy as much as possible with the building inspector: the expert who legally sanctions the house on the building site for the city. The desire was to see what processes occur when a layman presents a set of un-expert but mildly adequate plans. Such a set of plans represents a considerable saving over the hiring of an architect and having elaborate plans drawn up on order to immediately please an inspector. Elaborate plans are of no necessity for building the house since so much is standard construction—controlled by code—whether on paper or not. This was an effort at cost control, where quality would not be affected.

In "going cold" to the inspector's office with a plan that shows as little as possible about the house and with the viable anticipation of receiving a permit immediately, the layman is playing patsy to the fullest. In my case, the inspector immediately found several things lacking and wrong in the plan. This response plus the patsy's saying, "I do not know very much about getting a permit, what must I do? Can you help me?" shifted the burden of correction and completion on the building inspector. The building inspector, feeling in himself the strength of an expert facing a willing-to-learn layman, was quite willing to help. He then agreed to prepare for the next day a list of several items to be done to acquire the permit. This list, then, amounted to a contract with the city. If the patsy did what is specifically required, it would be difficult for the inspector to refuse a permit, though additional things could be asked for. Of these items only one cost money—a contour survey with surveyed plot plan location of the house—so the dollar savings in playing patsy was considerable. (The contour map, later, proved to be irrelevant to the problem it was to solve.)

It is conceivable that the patsy could have been forced to spend a sum of money on a full set of drafted plans complete with engineering specifications. (A shakedown.) But, in this building department, they do not do this to patsies, only to general contractors!

In order to help out, the building inspector, a former draftsman, asked if he could draw in the porch framing and the heating and electrical outlets for the patsy. I agreed, and he completed this aspect of the plans with pride, even transferring his marks to all three copies so regulations would be complied with. It was obvious that this other expertise had been unexpressed when he faced all the fine plans that came through his office. But, at points, this expression slipped into a redesigning which mostly the patsy had to firmly refuse with thanks. However, the accepted points put additional responsibility on the shoulders of the inspector for accepting the completed house. This was a hint of the *redesigning process* which subtly occurs in expert-layman relationships (it will be taken up at length in Chapters III and VII).

The building inspector also made an extended site visit to study and report to the patsy via his list what had been lacking for the permit. The inspector discovered that a non-adjacency certificate and an excavation permit were lacking, and the septic tank permit was outdated.

He clearly had put in cost free time for what an architect would have put in only for a commission.

It was clear to me that the process of getting a permit would have to be gone through many times in order to become expert enough to obtain it on the first try without problems. General contractors may be able to do this. It was also obvious that a patsy never could; so he may as well go in cold with partial plans and get a full list rather than spend the time trying to be complete at first try, and still be told what needs to be done.

When the list had been complied with and the data presented, the inspector's response was two-fold. He immediately found more minor details missing on the plans. A countering grimace of trivia on the face of the patsy led the inspector to make the corrections himself and say no more. In facing the completion of the tacit contract and having to issue the permit, which made the city liable for permitting this construction, the second response was. "Now let us see what we can get him for before we give him this permit." This is the standard, slightly legitimate shakedown—a way of picking up all possible fees from a patsy before releasing the permit. Generals could not support such squeezing. Plumbing and electrical fees were discovered to be omitted—they required separate sub-permits. The patsy was politely asked (forced) by the superintendent of grounds who happened by, to agree to fix a city drainage problem in front of the property. He was then handed the permit in exchange for the total building fee. The patsy is then released to start generalling his job, with the statement. "Do not forget to call me for the foundation inspection before you pour *the cement*." This was a reminder to the patsy that systematically looking over his shoulder during the project would be the building inspector—the city's expert. There would be an inspection for the foundation, the closing in of the walls (the plumbing, heating and electricity) and the finished product, before the gas and electricity could be hooked up to make the home livable.

In playing patsy, then, the layman did as well as the experts, the general contractor and the architect, paying less money for his initial jobs. Both he and they end up with a building permit however different their means.

Financing: When shopping around for a construction loan, thus comparatively bidding, lenders automatically emerge because of the structure of lending done by savings and loan associations. Some will say that they do not back owner-builders (it's too risky because they will not finish.) Others will say that they are all loaned out for a month, and to come back on a certain date. And those who accept the plans to review the loan request take a few weeks to study the proposal and allow the loan committee to make up their minds. These conditions generate the tendency to keep looking for the best loan in the shortest amount of time.

A crucial aspect, with respect to our study, in comparing lenders on construction loans is the scheduling of their release of monies. How often they release a payment and who pays the subcontractors will affect the patsy's working relationship with the subs. These arrangements affect when the sub gets his money, and when the patsy says a job is completed. Infrequently some lenders will release all monies to the patsy immediately, others mostly in three to five steps which are synchronized with construction. The subcontractors must understand these payments, to understand delays in their payments, and how to plan meeting their own expenses. Also, some lenders will prefer to pay subcontractors directly upon receipt of invoices and others will "trust" the patsy to pay them after each release. (The same trust goes for general contractors, depending on their reputation and bond.) These policies will obviously effect the patsy's use of payments as a strategy of control in power symmetry. He must choose his lender carefully since maximum control is secured with many releases of monies to himself, spreading out the power of distribution of payments to subcontractors.

Once he accepts a lender's bid, the patsy receives a loan commitment for a specified amount of time, which he may or may not need to get things together sufficiently to warrant signing up for the loan. Problems may still exist with say, design or obtaining the building permit or clearing title to the land so that the lender can have a clear first priority position.

Once the patsy has signed up for the loan, he feels a surging desire to get the project going and to get it finished. This arises from his having signed up for a big debt and accepted the risk of spending large sums of money with little of value to show for it, until the true value is brought out by the finished product. Then, more value will be in the house than the amount of the loan, and the investment will have proven itself. There is a point halfway through construction when more than half the loan has been advanced and still there is only a skeletal mess to show for it—and that is particularly scary for the patsy. At this midpoint, the patsy may clearly feel the need to hire a general to make sure the job is finished properly and the money not lost. Thus, his faltering in playing patsy and his having the option to turn the job over to an expert amounts to an ever present condition which varies in intensity at different stages of construction.

The Generalling Attitude: These feelings of faltering or of potentially failing bring out the patsy's necessity to have the proper temperamental resources in order to attempt generalling. He must be able to put on the act of generalling with confidence and without fear. If he becomes scared when facing experts he thereby may easily abdicate his control as the sub moves in to run the job *his* way. Equality of power symmetry is lost. The patsy need not fear nor feel embarassed over lack of knowledge; he must be prepared to glean it from subcontractors and continually to reassert his control by using it. This approach is quite different than the typical layman's accepting his know-nothing status as a reason to turn over near complete control of his problem to an expert. Further, the patsy must have the ability to put on the several attitudes and stances of a general contractor to maintain the balance of power symmetry sufficiently equal. He must act as if he will give the subs more jobs and refer them elsewhere if they do good jobs. He must be able to get angry with complete control, be exquisitely patient, engage in baiting as encouragement, withstand cajoling and pressure, and withhold requested payments. He also must be prepared to general in person and over the phone. Playing patsy, he must never forget the subtly current shifting of power symmetry and his temperamental resources for gaining, regaining and keeping an edge of control.

Location on the Job: In generalling the construction, the patsy locates himself in the division of labor of home construction. He has several options and should be clear on which he takes and to what degree he takes them. This location is based on three interrelated dimensions: the level of division of labor, the stage of construction, and his desire to do things himself.

There are typically four levels in the home construction division of labor: general contractor, subcontractor, foreman and crew or workman. The patsy is taking over the general level and may do so for the whole course of construction or only for one of its stages. Thus, he may hire a general contractor for either the foundation rough-in, or the finish stages. He may, say, wish only to general the finish stage. It is most common for patsies to general the last two stages which are the home construction *per se*, and leave the foundation work to a burly general contractor.

Taking over the general level puts the patsy in direct contact with subcontractors, but he can bypass them and hire a foreman or workman directly if he knows of some. Usually these people are moonlighting, and some virtually subcontract without licenses. This direct approach necessitates more insurance—the insurance which subcontractors carry for workmen. In fact, it is not uncommon for workmen who frequently hire out directly to represent that they are subcontractors with insurance when they are not subcontractors and do not have insurance. The patsy can then be fined by the state for not having workmen's compensation, and by the government for not paying withholding tax, which he had thought the workman-subcontractor was doing. The patsy, then, accepts these responsibilities when hiring direct workmen and screens his subs carefully.

Supervising workmen rather than subcontracting may require more expertise in guiding and coordinating the job, and the patsy has far less recourse when jobs go wrong. The sub is responsible—he stands behind his work—because of his license and contract and the Better Business Bureau Contractor's Exchange, etc. The workman merely gets an hourly wage and can quit with immunity if things go awry. In this monograph we locate the patsy as a general using subcontractors, *not* as playing subcontractor also. Dealing directly with workmen can save money, but the risk is in quality control. Before taking on a workman directly, it is best to know him and his work thoroughly.

The patsy also has the option of bypassing, as a general, the whole division of labor and doing a job himself. It is typical for generals to do their own rough carpentry or directly hire their own crew for this. They may also do one of their other pet jobs: say, the electrical work or plumbing, and then sub out the myriad of other jobs. The patsy himself can do what he likes and sub out the rest. In my case, the patsy did the rough carpentry with the help and tutelage of two experienced workmen. This was very helpful in keeping tabs on cost and quality and temporal control, since it is on the rough carpentry—the framing of the house—and its pace that all else is paced. In this way the patsy took the typical approach of general contractors to generalling—rough carpentry and subcontracting.

However similar this approach is to that of the general contractor, none the less there are important differences between this expert and the patsy. Obviously, the general contractor as an expert, has more experience, skills and tools. He has done more and is doing more construction. Thus, he is expert in bidding and running subcontractors, especially in handling elsewhereism. In competing for the sub's services at the right price and on time, the general usually has more interpersonal skills and has the power to outweigh the competitive balance with unknown others having given more business and prospectively giving more business than a patsy. Clearly, the patsy is bound to find it more difficult and somewhat more costly to keep a toehold in this system of bidding and using subcontractors.

Having experience and a license to prove it and, when necessary, the tools to back him up, the general moves in more quickly on the jobs than a patsy, and can do the jobs himself or sub them out. The patsy, in facing a job, first must take the time to inform himself for subbing it or for doing it himself, and if the latter he may have to buy or rent tools. (We shall have much to say of this *information gaining process* in the next section of this book.) The patsy seeks details for his particular job, while the general has under his belt more general details that guide him in the variety of different ways of doing this generic kind of job, e.g., the cabinets or flooring. Thus, the informed patsy is only a specific expert for this job, not a generalist, but that is sufficient.

The goals of patsy and general differ, and from the goals derive a different care and concern in generalling. The general contractor builds for a party or for speculation in pursuit of profit. Quality, costs and time are focused on profit. The sufficient pleasing of his customer or his potential buyer has to fall within an appropriate profit margin. The patsy, in contrast. builds to use for himself or as if he would. His house is a special object then and he takes great care and concern for achieving quality in its construction. The general's concern is in calculating costs, timing and making a sale.

These factors of experience, skill, information and goal generate different pacings of the construction. The patsy takes the time necessary to subcontract adequately in order to get information, and to control quality and costs. He can easily delay the temporal order to rethink details and design. The general paces on a different basis. The faster the home is built, the greater the profit and, of course, he has the experience and organization to completely plan ahead and build fast, even many houses at a time. For example, the difference in pacing was brought home by a general contractor's construction across the street from the house which I was building. The general started five weeks later and was beyond my stage of construction in a week and a half. However, the differences in concern and care were brought out when the general came to visit my house. The general remarked at its careful details, which he admitted he either ignored or routinized in his "spec" house. In questioning the costs and time involved in the details (for example, the type of skylights or the size of roof beams), he figured that to do similar things would erode his profit margin.

In sum, generalling his own job, for these many structural reasons, the patsy is far from being a general contractor. (Toward the end of this monograph we discuss the pressure the subcontractors exert on a patsy to go on generalling, to become a general contractor so that meanwhile they can keep getting jobs.) Chances of his becoming a general contractor are very small, but the chances of his success in generalling the building of his own home are virtually 100% since he just keeps at it until it is done as he wishes.

Seasons: When possible, a patsy chooses the winter for building his home or extensive remodeling. This is the season of least building, which means that costs for materials and subcontracting bids will be at their lowest and that elsewhereism is least burdensome. Material costs are low because inventories must be moved at some price to meet operating costs. Subcontractors' bids are low because they need some monies to keep their crews during slack months, so they will be ready with their experienced team for the Spring rise in building. Since building is slack in Winter, the demands of general contractors elsewhere on subcontractors are reduced. Thus, the patsy is not forever losing his sub in the middle of a job, thereby suffering consequent delays because of an inability to compete with a general who gives more business. During the slack season he gets better service and quality, and his temporal articulation of many subs on the job goes less awry since subs are less rushed.

The patsy who takes a long summer vacation to remodel his home is in for many surprises and delays. Elsewhereism, at its peak, erodes his carefully worked out schedule to be finished before he must return to his occupation. The patsy who feels fortunate in obtaining an inexpensive bid during the Summer, may find his work not done, since he has not paid enough for a place in the system during this season. Though he may not realize it, his job becomes a filler job between more lucrative ones. His sub is always drawn elsewhere. For example, one patsy paid for a "moving in" fee for a front loader tractor to do a five day job. However, because of elsewhereism during the Summer business high, the tractor was then moved out at no cost everyday and returned for only a few hours in the morning between 6:30 AM and 8:30 AM before the other subs started!

The plan of the book follows the course of home construction. In Part I, the comparative bidding system is discussed in three chapters which are ordered as topics occur when bidding. First, I discuss processes involved in inviting bids from subcontractors, the detailing the prospective job to them, and then choosing between subcontractors, and how subcontractors try to beat the bidding system so they do not have to compete so well.

In Part II, I discuss generalling the job once a bid is accepted and a subcontractor chosen. First, the temporal control of *a* job is analyzed, then the temporal control of *several* jobs, at different stages. During the course of construction some jobs are being bidded out while others are in process. Thus, while the plumber, electricians and heating people are finishing the rough-in work, the drywall contract is being bid out. Next, we consider quality control of a job in progress and the last chapter of this part deals with the problems and processes of securing closure on a job. In Part III, I conclude with a chapter in which I draw together some major generalizations from this work on the expert-layman relationship.

Part I

Comparative Bidding

The comparative bidding system, as I have said, is designed to allow a person (layman or expert) competitively to bid experts against each other so the person taking the bids may choose the best qualified expert for the best price to do work according to an acceptable schedule, (or in the case of bidding merchandise, money, commodities, etc., to choose the best quality and/or most quantity at the best price.) The prime function of this system regarding experts is *protection* from exploitation by experts, through giving the layman a comparative control over the interrelationships among price, quality and timing. In subcontracting, using the bidding system is the expected, legitimate procedure and provides the initial basis for equality in the power symmetry between the patsy and the subcontractor. Comparative bidding is one highly valued American way to achieve orderly competition.

In contrast, for general contractors, the prime function of the subcontracting bidding system is to find the best prices relative to quality and timing of the total job. In this way, a general can put together a total price for the house that leads to a profit, whether for speculation or custom building. He does not need protection from subs, being fully aware of decent prices and what to expect temporally regarding quality ranges from most local subcontractors.

The bidding system has two other important functions for the patsy. First, it provides a systematic occasion for engaging in the *information gaining* process, which is crucial for the inexperienced patsy vis-a-vis specific jobs and the general articulation of the total construction. This process allows the patsy to discover, as he interviews each subcontractor, which kind of jobs can be done and what kind of jobs he wants, and when he can ask for jobs through what he is told by various subs. As the patsy gains in experience this process becomes less important. (We lake up this form of learning process in Chapter III, on "*Detailing the Job*.")

Second, bidding and comparing provides the data upon which the patsy develops a resultant *trust* in his chosen subcontractor.

Concomitant to this trust is his effort to minimize risks through comparisons about the quality he will obtain for the price, and whether the *promised* timing can be depended on. For example, does an "economy bid" mean that the promised time of arrival on the job will be continually altered by elsewhereism—other people paying more for the same job—or that the promised duration of time on the job will be adhered to or will be enough to secure the desired quality? Comparisons take the patsy a long way toward making accurate judgments about trust and minimal risks associated with making adequate choice of a subcontractor. (Trust and choice are considered at length in Chapters III and IV) Under what conditions trust and comparative choosing, hence protection. break down is considered in Chapter IV, the section on how subcontractors beat the system.)

II

Inviting Bids

Bidding System Rules

To begin engaging in the subcontracting bidding system, the patsy need only know that he must ask several subcontractors for bids on jobs and must compare bids to find the best price relative to time and quality. Such rules of bidding that exist are few and flexible, and they emerge quickly as the patsy begins to invite bids. For example, he will quickly find out that bids are free by looking for subcontractors in the yellow pages of the phone book where "free estimates" are advertised. He also quickly finds out that bids are to be kept confidential. It is an impropriety to divulge one sub's bid to another, though it makes good leverage to get another to lower a bid. Clearly this confidentiality rule allows much fruitful deviance for both subcontractor and patsy (as I shall bring out in Chapter IV of Part I on "Beating the System").

The rules are not codified anywhere for the patsy's reviewal. They emerge quickly because they are generally known and adhered to tacitly. Breaches are clearly ignored or avoided. Consequently the patsy picks up his transgressions quickly when trying to keep his interaction with the subcontractor flowing toward his desired end. For example, subcontractors clearly avoid asking what others are bidding and, usually, avoid hearing any droplets of another sub's bid made by the patsy. Another rule of confidentiality is that one does not let two subcontractors bidding on the same job meet at the job itself. To be sure, that would be awkward, but more importantly it would contaminate the competitive bidding, if they knew, for sure, whom they were bidding against. Knowing each other's relative price ranges, they could adjust bids accordingly, instead of giving their best job for the best competitive price in a generalized market. Thus, if the novitiate patsy engages in the impropriety of having two subs come to the site at the same time, then one will surreptitiously wait in his truck until the first sub has left. While waiting he may feign not knowing who his rival is—though he cannot but know if the rival's name is painted all over his truck.

The emergence of this subcontracting bidding system is virtually unobservable to any other person. All know it is going on, but none can really see it happening, save for such an impropriety as just mentioned. The patsy and subcontractor relationship is private and confidential at this stage. Even the process of how and whom the patsy finally chooses is completely invisible and is unaccountable to anyone else. The patsy can of course tell anyone of his bidding experiences and determination, even seek consultation, but he need not—and when he does it is *ad hoc*, relative to the system. The slight crack that it poses to the system is soon covered over. The privacy of the system protects both the patsy and the subcontractor from annoying intrusions into their mutual deal.

The system's flexibility and invisibility puts its control almost completely on a morality basis; it also makes for individual strategy as on a calculative basis. There are no legal contextual controls (save general criminal and civil codes) and infrequent and little informal, concerted social control. The latter may emerge, say, when plumbers get together at a bar for a beer after work. If one mentions that a patsy is mickey mousing his bid, others may replicate his experience or simply pitch in to help put the patsy in his place. Then, the patsy *may* find himself controlled by all reputable local plumbers. More likely, one of these subs will use the data to beat the system and get the job. Once out, the word on a patsy can travel fast, and interpretations about him will vary very much.

Various bidding systems may differ considerably with regard to their rules of procedure, flexibility, observability and control. In contrast to the subcontracting bidding system—which is flexible, nonobservable, nonaccountable and confidential—there are several quite visible and rigidly controlled institutionalized bidding systems: as in large-scale general contracting commodity markets, and in auction and oil and gas leasing markets. In these, the opening and timing of bidding to the public may be publicly announced or advertised and the rules firmly set forth. To maintain secrecy and control, laws provide for sealed bids, neutral panels to evaluate them and/or clear evaluation criteria. Hand bidding signals are clear, and bids always indicate a clear legal commitment, if accepted, for the bidder to come up with the work or money on *time*—or he will be legally penalized plus, perhaps, lose a purchase or a job. By contrast, patsies will seldom if ever penalize a subcontractor who does not comply with *promised time*. But large-scale general contractors may be penalized thousands of dollars for each day that they are late with a satisfactory completion. Bidding systems (like subcontracting) will be more flexible and less controlled when what happens as a result of bidding is least pervasively fateful to the public, the communities of bidders and the takers of bids.

This flexibility of subcontracting also may be contrasted with the rigidity of professional experts (such as doctors, lawyers and accountants) who permit no concerted comparative bidding. Their initial strategy to avoid comparison is to charge for the first visit and immediately start to act as the professional on the job or refer the client to a more appropriate professional. When referring, even though ethics may require mentioning three people, the expert may sometimes direct the layman to at least one specific referral implicitly—thus comparatively pointing out the best expert before the layman has a chance to comparatively bid them himself. True, the expert then is saving the layman money and time. However, in subcontracting the patsy makes his own decision as based on first-hand data.

Now let us turn to discussions of: 1) how the patsy initially starts to protect himself when inviting bids, 2) his initial call, 3) how he arranges the site visit of subcontractors during each stage of the construction (sub-floor, rough-in and close-in periods), and, 4) how he handles the "can't say no" pattern of subcontractors—a prevalent pattern among many kinds of experts.

Protection

The Referral System: The patsy begins by inviting "free estimates." As all laymen know, the protection obtained against incompetent experts will vary with the kinds of sources by which one finds them. Thus, the prudent patsy goes to sources that initially increase the probability of protection against scoundrels and generate trust in "good workmen." There are three types of sources that he may use: being referred to, direct seeking of subcontractors, and solicitation by subcontractors. Within each type he seeks the most validating source and each one is related to the general referral system of subcontractors.

Regarding referrals, he will ask friends who can refer someone, "Vouching for" from experience. Also he may ask for an approved list from the local contractors' exchange, better business bureau or related business people such as general contractors, material suppliers and salesmen, other subcontractors in a different field, or a realty firm.

When he starts asking around for "good" subs, he soon finds himself in the middle of an intense, willing and affable referral system. As in medicine and law, it is not hard to find out who "the best men" are; their names are on everyone's lips and list. The intense part of the referral system is the ready referral by subs on the job of subs in other fields. The former do it mainly because of probable reciprocal referring—job feedback: "Tell him I said to call." Sometimes they do it for a rebate. And sometimes they refer another sub in their own area as a polite way of disqualifying themselves just now because of the job's size or because they have a full work load elsewhere: but they do not want to cut the patsy off without placing him in someone's trust. In this way the referring sub hopes to be called back later for another job, perhaps when he needs work. Or if he does not want a callback, still he does not want a negative image of himself floating back into the referral structure because this will damage chances for future, more appropriate calls from elsewhere.

However, there is a risk in giving referrals. If the referred sub gets the job but does poor work, it reflects back on the referring sub as one who cannot be trusted. With patsies the danger is less, since they do not discern poor quality and timing as well as do experienced laymen or general contractors. But the danger is still there and many a sub will refuse to refer—having discovered that it never works out well on feedback or having lost customers through poor jobs or even having seen the referred sub in turn referring other patsies elsewhere but not back to him.

Other subs may be so eager to refer that they eavesdrop on the discussion of a job, and then break into a conversation with a recommendation of a "good" sub. Or they may leave business cards all over the floors for the subs who will do the next job after theirs. For example, the dry-wall contractor drops cards for a painter. Material suppliers will refer subcontractors in the hope then of being used more by the sub. One lumber yard did this and broke the "yellow pages" (business phone numbers) monopoly of a door contractor. The lumberman hoped to sell wood to both subs, since the latter, not being in the phonebook, lacked business. Another supplier of glass referred a patsy to its installation division. This amounted to a referral-solicitation. Subs who are out of work often travel as a crew with subs who are getting jobs, in order to weasel implicit referrals. The sub proudly shows his crew as another sub, while the latter suggests calling him directly next time.

The intense part of the referral system also produces strong cross validating references because many referrers lean heavily on "good" or "best" characterizations when referring. As these cross validations pile up, trust and protection in a few subs grow to the point where the patsy knows he will be comparing the bids of a competent range of reputable subcontractors.

The patsy is fed into the referral system via his search for subcontractors through: the yellow pages of the phone book, signs and phone numbers of trucks seen in the area, hearsay on subs from friends, and through friends who happen to be subs. The yellow pages present an initial array of comparative data by which to choose which subcontractors to invite. Besides advertising free estimates, the advertisor may describe the size and type of job specialized in and their firm's history ("since 1954") in the county—which survival validates, presumably, competence and guarantee. The advertisors may also advertise how fast they work and how accessible they are ("24 hour service"). After seeking a few subs, the patsy then gets referred around to others in other fields.

When seeking friends who are subcontractors themselves, protection is maximized or minimized depending on whether or not the patsy knows the friend's work. If it is quality work, then o.k., but if unknown and the bidding does not work out, it may be hard to refuse the friend solely on comparative bidding grounds. Friendship is a condition that breaks through the bidding system and erodes the patsy's protection. (In two cases the author suffered by not refusing friends the jobs; he had to support delayed and frustrating work situations along with mediocre quality.) Also a search is required when as a cost control measure, the patsy wishes to find workmen who are not licensed but will subcontract for less. Often they are only found outside of the referral system, so they are not legitimated by references. Finding them is sensible for very simple jobs, such as tree cutting where complexity and quality do not warrant higher costs. Another type of direct search consists of callbacks to subcontractors known through other jobs or experience.

Another aspect of the referral system is the solicitation of bids by subcontractors. In essence, most soliciting is based on someone's referral to the sub telling him to call the patsy—which is the alternative to the patsy calling the referred sub. Thus when the patsy's building permit is approved, the city notifies the builder's exchange which publishes a newsletter or paper on who is building current projects. The paper provides those subcontractors who need work with the names and numbers of whom to call to request permission to give a bid. Again material suppliers or other subcontractors may tell subs whom to solicit "cold," or ask the patsy if they can refer him to a few subs so that the latter can solicit—thus bringing the patsy into the referral system. Some contractors as a preliminary solicitation send out letters about themselves and follow through either cold or with permission. Touring is another form of solicitation: a sub out of work may travel through a neighborhood that has a lot of building, asking the boss at each job if he can bid on, say, "the electrical work." The overflow from other jobs close by also results in soliciting. Thus the sub is called to a project down the street and after tendering his bid, drops in on the patsy to ask if he can give a bid since he is working in the area. These unarranged site-solicitations either can be embarrassing if other subs are bidding the same job on the site or annoying and interrupting of ongoing work. They seldom work out. The fastest strategy for removing the intruder is to thank him, say the job is taken and promise to refer him elsewhere: in short, the sub is placed back under the controls of the referral system which he tried to get around. When solicited bids look good, however, the patsy can turn to the referral system for validation references merely by checking around to see who knows the soliciting subcontractor and his work.

Types of Subcontractors and Fit: Through the referral system and the accompanying validation process, the patsy also finds out rather quickly that there are various types of subcontractors. He finds which to ask, and which afford the most protection by finding the best "fit" between sub and job. Some subcontractors only do new construction or large remodeling jobs, while others repair and do small alterations. Some do both. The patsy conserves his time by learning to call only those who do new construction in his own "proportion." That is, some subcontracting firms are so large they wish only to do a number of houses at a time, and some are so small that his size of house can be a burden. A clear way to find out what type of sub to call is through the business pages of the phone book. Many subcontractors advertise the type of construction and the size of job in which they specialize, also their own size. They may also advertise the location where they do most work: "serving all X county" or "serving north county." The referral system also specifies these dimensions when the patsy is passed on to a sub who fits his needs. What the patsy does not know about the sub can easily be found out by an initial phone call designed to discover if enough of a fit exists to warrant a visit. (I shall discuss the initial phone call in the next section.)

These basic dimensions of subcontractors configurate to give the patsy a notion of the right type of subcontractor from whom to invite bids in order to gain power symmetry over the working relationship and to maximize protection of quality, costs and temporal control. For example, if the contractor is too large or too small, temporal control is easily lost to elsewhereism. With the large contractor, the patsy can never hope to compete with the sub's many large clients—who can demand immediate attention. With the very small contractor, the patsy will find that the sub is "in and out" of business in order to stay alive: he can easily be drawn elsewhere if he has to work for another big contractor for a week or two or more to make a living, especially when he does not have enough current jobs.

Quality control may also be in question with the large and the very small subcontractor. The large ones work fast using a work hierarchy in depth and once they consider a job done they can be immune to the patsy's maneuver, then back again to do the job right. For them, one small liability does not hurt. On the other hand, the very small contractor has not the financial backing to stand behind his work: his guarantees do not carry weight. For "system" type jobs—like plumbing, electricity and heating—a guarantee and the ability to back them up are vital.

Also, the territory where the subcontractor does most of his work is crucial. If, for need of a job, one goes out of his territory, the patsy then may discover that his quality and temporal control is poor. Thus, the contractor may be delayed making his visits, for these are interruptions to normal rounds done in another territory. And when out of it he may work too fast in order to get back home, thus jeopardizing quality.

Site-Bidding: Finding subcontractors proportional to the new construction is not difficult, as the too large and the very small contractors are in the minority. Most take separate single-family residences as their regular business unit. Most also engage in site-visit bidding which is crucial for the patsy's protection. (The two primary paths to bidding a job are on the basis of plans or site visits.) With the site visit, the sub for complex jobs may also request plans to supplement his site data with specifications ("specs"): these are the "system" jobs of electricity, plumbing, and heating. For a few simple, quick jobs—e.g. tree removal, or a flat garage roof—the patsy can phone the bid. The job is so simple to detail that it can be handled over the phone without mistakes once the patsy's detailings are trusted.

The knowledgeable or prudent patsy will want to do all site-visit bidding because it maximizes control over quality, costs and timing. The site-visit bid is as accurate to the facts as a bid can be. The sub sees *en situ* exact measurements, possible obstructions, exact material needs, timing to and on the job. Thus he can provide, for the best costs, the desired quality and timing. Plan bidding is less accurate because it does not take into account the site problems and because of the subtle changes in plans that emerge in single home construction whether custom built or by a patsy. To protect himself against going over the bid amount for the unforeseen, the subcontractor puts in a "cushion" involving costs and clauses that make changes or additional "extras" in costs. Thus the patsy cannot be sure that the large bid will even be the most he ends up paying. Plan bids allow miniscule "gouging" of the patsy. The plans themselves cost most than those necessary for a permit and site bidder. Plan bids are designed for large housing developments and big buildings, where bids are needed before building and where site visits are too time consuming. When general contractors bid subs from plans for a total on custom or speculation houses bid, they get estimates (or guesstimates) which are sharpened (down) after the house is begun.

Unlike plan bidding, site visit bidding requires temporal pacing of the job. A given job cannot be bid until enough building structure is present for accurate detailing, but the site visit must be done long enough in advance so that the bidding will be over and the subcontractor ready to start immediately when his job is required to move the project along. Thus, poorly timed site-visit bidding may delay the total project and other subs, or if done hastily or forced may compromise cost and quality on jobs that count—such as visible finish work like tiles or floors. Such delays and compromises can have a cumulative effect, as subs get involved elsewhere when they cannot start a promised job on time and then cannot get back to the patsy's job when he wants them. But these dangers are much less than the unexpected ones that come from plan bidding.

For the patsy other benefits accrue from site bidding as over plan bidding. In detailing the job, the patsy can specify exactly the most minute details. He can, with the sub, redesign aspects of the job on the basis of emergent realizations of taste and convenience and of new information. In front of the actual job, he can develop for the sub feelings of trust and risk on timing and quality. (For example, the sub who continually redesigns the patsy out of his wishes, undermines a trusting relationship. He becomes suspect because he may possibly make unauthorized changes.) In the end, the patsy can obtain the most complete contract and proceed temporally and budget-wise with confidence. On the contrary, the benefits of solving all detailed problems beforehand constitutes deficits that turn plan bidding into a morass of uncertainty involving possible delays and higher costs: once the job starts, the subcontractor finds what he is really facing.

Number of Invites: For adequate protection, it is normal to invite three bids on a job. Two bids do not allow a range of bracketing. With three bids, the high and low are indicated by a middle bid. Also, by the third bid the patsy can pretty much tell what quality he is going to get for what price and whether what the job is typically going for "these days" is worth the money. Four bids lead to getting the same things over again and becomes tiring.

If, because of his budget, the patsy is seeking a not too risky cheap bid or a sensible bid for a special job, he must really search. Prices tend to go up for special jobs, since regular prices are worked out for the standard ones. Special jobs mean decisions and delays for subcontractors which cost them more. But there are subs who are used to specialty work and keep the costs of a job down. Sometimes a subcontractor will not bid on the special job that has too many unknowns and will only work on time and material, so he will not be penalized for delays. Thus, one masonry contractor has a regular price for the standard fireplace, but will work only on time and material for "fancy jobs" where "the brick shows everywhere."

Bidding contractors can easily sap the energy of the patsy. An unending stream of subs constantly going over details is so exhausting that it becomes easy for the patsy to give up bidding on a job—hence giving up his protection—to save himself time and energy. It is particularly tempting when the first or second sub indicates a complete grasp of the job and immediately wins the patsy's confidence. At these points the patsy either must force himself to invite more bids or take the risk of receiving less protection. Indeed, if a sub should discover that the patsy is not bidding out the job, he is likely to add a little something to his price. Thus, the fatigued but confident patsy who foregoes bidding protection must keep this secret from the sub. He will put on an air of having chosen the best of many bids.

For the inexperienced patsy, his only initial protection is bidding. Later he may develop other ways to protect himself. He may meet other patsies whom he can trust; for they have bid a job and can tell him the best subs. Also, he becomes more experienced in judging subcontractors. And as he develops strategies to compare them to general conditions—not with specific others—he may not comparatively bid out a particular job because he believes that he can protect himself with just one bid. He may, for example, have an idea of the range of prices which he would pay for laying a floor with a particular material that he has chosen and priced out. The job is simple enough not to make quality and timing a large risk—it can be done in a day. Thus, at the finishing stage he may skip bidding a simple job, of course keeping his sub unaware of the lack of competition. Patsies still bid out complex, time-involved jobs such as ceramic tile work when judgments are too hard to make without specific comparisons.

With reference to home repairs, other reasons that patsies abdicate the protection of the bidding system are delineated in the following quip from the *Wall Street Journal:*

> An increasing number of Americans seems to be avoiding the difficulties of bid-taking by eschewing the process altogether, specialists in consumer affairs say. They say that some buyers are too timid or feel they lack the knowhow to evaluate cost estimates in technical areas. Other buyers apparently feel that bid-taking is beneath them. "There's a growing class of newly affluent persons who have come to regard getting bids as 'nitpicking' about costs," says Robert Smith, assistant director of Consumers Union, a nonprofit organization that publishes buying guides.
>
> Whatever the reason, it's clear that people who don't bother to get price estimates before they contract for service work *wind up paying more.*

The difficulties in bid-taking cannot be gainsaid, but experience makes it easier. The need both for the temperamental resources and for yielding the attitude of turning "'all" problems over to the expert has been discussed earlier. The author in the context of this study did not run into upper social class or cultural definitions that prevented people from subcontracting. Many people and subs also said they found bargaining distasteful, but none proved so affluent relative to the large sums involved in home building to offset a try at price control through bidding. These patsies, though, were easier for subs to overcome by making slight accusations that they were universally "arguing over costs" when the vital issue of "class of their home" was at stake.

The Initial Call

The initial call to invite a bid poses the problems of reaching the subcontractor or his representative, and then briefly establishing sufficient fit between his type of work and the patsy's job. Only with potential fit does it make sense for the sub to take the precious time for a site visit bidding.

Upon making the first call the patsy will either reach the sub and start establishing fit or have to leave a message for him to call back. Let us consider first the message phenomenon, and then the establishing of fit when in direct contact with the contractor.

Messages: The chances of reaching a subcontractor during the work day is, at the outset, small. Subcontractors usually are elsewhere—out on the job, isolated from potential clients while working with other clients. Thus, the initial call results in the patsy running into several different message takers, each of whom must be handled somewhat differently either to secure a "call back" or to find out when he should call back.

The patsy may leave a message with a secretary, answering service, wife, a crew member hanging around the office or make a recording of a phone message. The recording promises a callback, though it may never occur, in the voice of the sub himself. If a woman is talking, the patsy may not know whether she is the secretary, wife or answering service. Having been asked, she must be handled differently with respect to giving and receiving information. The answering service will take a perfunctory callback request, while indicating her ignorance about such subcontracting conditions as timing. The secretary and wife may engage in a very brief initial screening for fit before taking the request. Their screening is usually based only on whether the job requires new construction or repairs. The patsy may "bait" a little to ensure the call back by indicating who referred him to the sub, or saying he requires new construction. (When answering services screen for doctors or lawyers, they do so on the basis of emergency criteria; this only happens for subcontractors involved in emergency business—e.g. plumbers.)

The secretary screens to protect her boss from needless work and clients who are trying any ruse to get ahold of him to finish, correct or otherwise get back on a job. The wife screens for these reasons and more. She is vitally interested in her husband's business, its financial side, and his time free from business for alternative activities like meals, showers, night outs and leisure activities. The wife works for nothing, yet for her own life, while the secretary gets a salary; neither is allowed to let a client get away, but of course the wife has more to lose (financially and emotionally) so works harder at screening and ensuring a callback. The wife will be more careful about arranging who will call back and what are the best hours for the patsy to give or receive a call. Yet often the wife is more in the dark as to her husband's schedule than is the secretary, who is on top of the daily office schedule. Like answering services and secretaries, some wives are very alert as co-partners with their husband; others are slow and unaware, and try as they may, will botch the message. Some wives purposefully botch it and pour over the phone their anger at their husbands. In these cases, it is better for the patsy to just keep trying to reach the sub.

The promised callback from all message takers is typically linked with a more or less definite promised time—which occurs "as soon as possible," "when he gets in," "the following afternoon, evening or morning." The patsy need never wait more than 24 hours; for, he depends on this promised time for getting the bidding on a job over with. The message phenomenon is crucial in all expert-layman relationships. With doctors it can be a matter of great pain or even death; with subcontractors, the patsy soon learns to keep calling others when callbacks fail to occur for bidding. And when a sub has the job, the patsy learns to ensure his callbacks by timing the sub through his treks elsewhere back to the office to maximize the certainty of reaching him.

Several conditions determine who will call back whom. The subcontractor may call back, as we have just indicated, as a matter of policy. The policy derives from scheduling his time between making phone calls and working elsewhere. The message taker may leave it up to the patsy. If it concerns the third or fourth bid, the patsy may just leave his name and let the subcontractor call back at his will, making no arrangements for it.

The patsy may call back because he needs immediate bids or a quick job, whether or not the sub is supposed to call back himself. He may call back because he believes the message taker (the slow wife, the harried answering service) could not handle the message or because he is invited to call at again a set hour. If the patsy detects reluctance or disinterest, even with baiting, in the callback arrangement by the secretary or wife, he suspects or it becomes obvious that the sub is too busy elsewhere to need his job. The sub, therefore, is not likely to call back, and the patsy must press for the bid by calling back himself. In this event, he loses some edge in eventual power symmetry if the bidding proceeds. At this initial stage it is not unusual for the patsy to make several temporally "braketing-in" calls before reaching the subcontractor when the callback is not arranged. As a last resort the sub will be home at 6:00 AM!

Establishing Fit: The subcontractor is reached through: a direct contact; a callback; wading through secretaries, hierarchies and departments; or waiting out the sub who is on "another line" talking elsewhere during his phone call hours. The patsy, then, proceeds as fast and briefly as possible to establish a preliminary fit of job to subcontractors. It is typical of expert-layman relationships to spend as little time as possible, usually over a phone call, to see if a preliminary fit warrants a personal meeting over a possible extended relationship. Establishing preliminary fit occurs on three dimensions: type, location and details of job. Efforts will be made by both parties to review these dimensions in that order; but if, say, job details happen to be mentioned first, comments by the sub are suspended until location and type of job are ascertained. During each phase each may try to "bait" the other toward assuring a site visit.

No matter what the outcome of this effort, the patsy will invariably run into "sweet talk" by the subcontractor or his estimator. Subcontractors make a practice of always sounding like the "nicest guys in the world." Agreeableness and compatibility is a rule that they all follow so as not to jeopardize, first of all, their position in the referral system and callbacks at later dates if they decide now to refuse (usually by referring a more fitting subcontractor or a less busy one). Also they sweet talk to start gaining the trust that they know is a crucial element in having their bid chosen. From gaining trust in the subcontractor, the patsy also gains the feeling that power symmetry is sufficiently equal to maintain quality, cost and temporal control. However, the patsy may lose the protection of the bidding system by being taken in by an exceptionally sweet talker.

Several elements go into making known the type of job. The patsy soon learns from contractors the questions to open his call with: "Are you interested in bidding on new residential construction?" This statement immediately establishes the fit between type of contractor and type of job, and perhaps becomes the basis for a continuing negotiated relationship. This is enough information for many subs immediately to say "Sure, where are you located?" or "Sure, may I have your name and address?" and no further fitting process is necessary; the site visit is merely arranged. Yet many other subcontractors will want to know more about the type of job.

The next element to establish is that one wants a site-visit, not a plans bidding job, which may but usually does not disqualify big contractors. Only, they will want the plans as a supplementary detail. Site-visit bidding makes sense even to a plans bidder as soon as he discovers through rapid questioning that he is dealing with an owner-builder, not a general contractor. The subcontractor then also wishes to weed out discrepancies between plans and actuality, so as to reduce possible future trouble. He does not trust the patsy as he would a general contractor to not stray from the plans.

A further aspect of the job that is of immediate interest to subs is the financing arrangement. Some will not touch a job, especially a patsy, unless there is a proper institutional building loan assuring their remuneration. The patsy after a few calls, learns to mention his building loan as baiting to get subcontractors to make a bid, knowing that many carry large unpaid accounts and so look for jobs where money is assured on schedule.

The patsy may also learn to use as an opener the bait of who referred him to the subcontractor. "Hello, X contractor (a plumber) referred you to me as an excellent mason for new construction." This phrasing very likely throws the full weight of the referral system behind the patsy's request for a bid, barring a grossly disqualifying lack of it. The subcontractor immediately registers to himself the feedback on referrals he has made to other subs and cannot refuse taking care of another, but different type of subcontractor's customers. That would jeopardize his place in the system. This acceptance can be nudged even more by the patsy who adds that the referring subcontractor is a "good friend"—which indicates that feedback will reach the latter as to how the sub responded to the bidding request. Subs do not like to refer subs who give a poor showing as this negatively reflects on their judgment and expertise. They wish to get clients into a good deal, not into trouble, so that they themselves will be used for future work. The same feedback of the referral system occurs when the referrer is another layman, though not as strong as if he were a contractor. Such baiting strategies with sub referrals can backfire when the referring agent is a sub who has lost his credit. The patsy may also become suspect or plainly ignored. The counter-strategy, if the patsy discovers his error, is to mention the building loan to establish his credit.

Another opening strategy is for the patsy to refer the subcontractor to the "builder's guide" where his own construction is listed, as verification of his story. Often the subcontractor will remark that he remembers seeing the patsy's name in the guide (a positive reference) or that other contractors had mentioned the patsy was building a house (word being out with the referral system to many subs who might expect a call to bid). Again, these opening strategies are important because a false site visit might waste a full morning for a sub, a waste he can ill afford since many visits do not pan out anyway and many bids are rejected.

As soon as the patsy opens his invitation by noting the type of job, he can virtually hear in the voice of the subcontractor the degree of elsewhereism which he is facing. If the sub is eager, elsewhereism is not yet a problem for the patsy. But any hesitancy will signal that it is: thus before going on with the fitting process, the patsy may offset some of the competency pressures by standard baiting strategies. He may mention the building project across the street which needs a bid on the same type job, so the sub can get in two bids during a single trip. (This strategy worked well for the author until the house across the street got too far ahead of his own.) He may demand the bid fast as part of the pacing of construction, requiring the job NOW so as not to hold others up—jobs such as getting the sheet rock up or having the roof on so that finish jobs can start. This strategy can be followed through with "I have several bids already and am about ready to choose so I must have your bid soon," or "Don't let the pigeons fly away." As with most efforts to combat elsewhereism, the details of which are completely unknown, the patsy is fishing in the dark. He knows that he succeeds only by the behaviorial indicator that the sub agrees to a site visit and to submit a bid.

Once sufficient fit on the type of job and necessary baiting are established, then either the subwill refuse to come for lack of fit, he will come with no more questions, or some brief details of the actual job may be required to ascertain additional fit. The patsy's brief detailing of a job allows the sub to size up whether the job really is his type of operation. There may be aspects of the job that he does not do and this disqualifies him. For example, some foundation contractors just pour steel and concrete and others make the forms as well; some paving contractors do not make catch basins.

This detailing by the patsy usually amounts to a succinct paragraph, perhaps followed by a few questions from the subcontractor.

> "I have a 20 × 30 carport roof that needs to be tarred and graveled. I just want gravel stops around the edge and there is a 4 × 3 skylight in the middle that I want flashed with gravel stop."

The sub's follow-up question may be, "Can I park my truck right next to the garage?" Or the patsy may be asked for a rundown on the course of construction to see if he is actually *ready* for a sub, e.g., for formica on cabinets or the plumbing rough-in. If the patsy does not offer these detailings he may be asked for them. As the patsy invites more bids on the same job and as he gets further into the project, he becomes sharper in his brief detailing and offers it immediately without being asked, after ascertaining a fit on the type of job. This brevity cuts down on time and effort in bidding. This brief detailing also may result in "no fit." If a fit is probable, it is likely to generate a need for the contractor to come up and see the site—"I better come and look at it"—for himself, when the patsy is ready. Thus a site visit is arranged for some future date.

If the job is simple enough (e.g., the garage roof) and the patsy clear and trusted, the brief detailing may be sufficient information and the subcontractor will give his price over the phone. Thus, the patsy finds himself phone-call bidding. The patsy then may go over further details, such as with the small roof, deciding on the kind of gravel and number of tar paper layers.

Other conditions may cause the patsy to try phone-call bidding along with brief detailing. He may try to quicken jobs; then he must establish enough trust to get a price without the sub having to worry over unforeseen conditions that may jeopardize his profit. The sub may also have already seen the house on his regular rounds. He may check on the patsy's reliability of detailing by calling a sub from a previous job. Most subs will decline phone-call bidding except on the simplest jobs—such as tree cutting, or paving a driveway or path, or hanging a glass shower door. Phone-call bidding may also occur because the property is too far away for checking out the simple details given over the phone. It would cost the sub more to make the visit than to risk a slight bit of his profit. The sub's being drawn elsewhere to a bigger job also may lead him to risk a little of his profit via a phone bid. For example, a simple tile job to be done on a standard bathtub will not draw him away from other more important bids or jobs; after checking for the bath's standard dimensions and agreeing to a standard tile he gives his standard price and time. Needless to say, phone-call bidding when it proves feasible provides considerable saving of time for both patsy and subcontractor.

Brief detailing may also include some statement of price ranges by either patsy or sub to see if "we're talking something meaningful." When the sub is "out of sight" with his prices or the patsy cannot afford the normal price range, then lack of fit makes a site visit worthless. When their respective price ranges overlap or are agreeable to the patsy, this leads to the site visit. "Well, Mrs. X, it sounds like you're talking somewhere around $175 or $200." If agreeable, the patsy says "good," but if not and the price is still close to a feasible range, then the patsy may say "we are getting somewhat better bids, perhaps you should look at the job?" This baiting strategy is likely to result in a revision of the price range plus a site visit. Another strategy is to ask the sub to explain his figuring, "what is your price per square foot?" to be sure that he is not building in a cushion that will dissolve after he views the job, or to be sure that he has the details straight. Sometimes the patsy also can note that part of the job will require materials that the sub later can make a resale profit on—another possible bait to keep the total price within the range that will make a site visit reasonable.

The last step in the preliminary fitting-process is to see if the site's location fits the sub's current business requirements. Here elsewhereism comes into play. If the sub is not busy, he may go anywhere (even to other counties) just to keep work coming in. But if he starts accumulating clients at one end of the county, he may resist going to another part for just one job. He may purposefully cluster his work in two areas to keep his crews hopping quickly between jobs. Thus, the elsewhereism generates boundary conditions pertaining to where the sub will "gladly" work (as is typical with all experts). Thus the patsy may be disqualified on locational grounds: the sub says "sweetly", "We are not figuring anything down there. Mr. E wants to sort of try to keep close because we're short of plumbers." The patsy has no way of predicting these geographical boundaries in advance. In one case, a patsy went out of his county for bids in order to break the local monopoly of a door contractor. He found that only big contractors and small ones without work made far away site visits, and then usually on only one or two days a week when enough calls had accumulated. Ordinarily this, of course, would slow down the pace of bidding and threaten the construction pace, so these calls would be avoided. For local subs, the patsy could demand immediacy in bids and work; however, the one out-of-county contractor never did arrive because not enough work and deliveries developed by the promised date to warrant the trip to the patsy's site.

If the general location is feasible, then the subcontractor begins getting directions. As directions pour out, if he finds the house is still too "out of the way" or requiring crossing roads difficult for his equipment, then he may warn of extra costs and so either re-establish a new price fit with the patsy or disqualify himself. Directions are important because the subcontractor dare not get lost, as that wastes time which he will spend at no charge. Being lost can cost quite a bit of money as also he may get fewer bids out that day. Thus, the patsy finds himself going over the most minute details of how to get to his site. Few subs are so eager to come that they say "Don't worry, I can find it," without getting pinpointed directions or without knowing the area quite well. "I'm doing a lot of houses up there." Once lost, a sub may easily give up looking in order to go on to the next bidding and to quit wasting time. The patsy's bid, then, is lost elsewhere.

Another locational dimension of importance to many jobs is the placement of the house on the lot. The most difficult placement is to go to the house up a steep lot. Special equipment may be necessary—for unloading sheet rock, bricks or roofing material—which the sub may not have or wish to bother with, or for whose use he may have to charge extra. "You have to walk up a hill, huh?" is the sub's pensive response. On the other hand, easy locations where a truck may virtually drive right into the house, bring lower prices and an "easy" attitude, perhaps even permit bidding via a phone call.

Once this last aspect of the fitting process is concluded, temporal arrangements for the actual site visit are made.

Arranging the Site Visit

The prime determinant for the timing of site visits is the pace of construction which requires that a number of jobs must be done. Mostly the patsy can foresee enough to arrange for bidding to be done without too much rushing. Infrequently, he must bid a job quickly—say, within a few days to get the roof on before it rains. When the "next job looms up" in the pace of construction, a set of bidding of subcontractors then emerges within a given period of time. One of the patsy's problems is when to invite this set of subs to his property without them meeting each other. In so doing, the patsy quite often engages in a contest between his requirements for temporal ordering of the visit—based on not delaying the pace of construction (these other subs') and not allowing subs to meet—and the invited subcontractor's requirements for being elsewhere at certain times. The initial balance of power symmetry emerges as a conclusion to this contest between the patsy's commitment and the requirements of elsewhereism. The contest does not occur when elsewhereism does not exist—for whatever specific conditions—so that the subcontractor can come at the patsy's convenience and/or the pace of construction is not pushing the patsy toward the next job. Flexibility about arranging the site visit will depend on the degree of the contest for control over the timing of the site visit. Let us consider the four conditions of: when elsewhereism is strong and pace of construction allows few delays, when only one condition is dominating, and when both conditions are relatively weak.

Intense Pace vs. Intense Elsewhereism: The contest for arranging the time of the site visit is most intense and least flexible when the subcontractor is busy juggling people elsewhere while the patsy is under pressure to keep the construction going so as not to delay subsequent subcontractors. To delay other subs and thereby potentially to lose them—which means further delays in construction—can result in a loss of control over the articulation of subs on the project and a loss of power symmetry in the relationship with each sub.

Under these conditions the patsy presses for his site visit as soon as possible in spite of elsewhere claims. The standard strategy is his incessant insistence on when the sub can and will come. After repeating the time over and over, the patsy ends on a promissory note to which the sub agrees as a definite *promised* time, ("You'll definitely be here tomorrow"). In his incessant insistence the patsy also continually repeats that he wishes to get the bidding over with, thus using the strategy that the sub will be too late if he does not come as promised. The patsy may also use the strategy of providing durations of time for the sub to come: usually a whole day or a morning or afternoon ("Well be here all day") depending upon when other subs are coming, and agreeing at the sub's convenience to be on the site during off-times such as weekends or after work when elsewhereism does not plague the sub so much. The patsy also may reiterate how large his job is, in order to compete with claims for site visits elsewhere in trying to snag a sub within a day. (For example, this tactic broke through the glazing contractor's week-long tight schedule, since he did many jobs which were usually smaller than that for new construction. "Next Monday" became "tomorrow" (Wednesday) almost imperceptibly. The clustering of all these strategies was necessary to break through the paving contractor's weekly schedule and to get a definite promised time within one day.

Other strategies easily emerge from the situtation. One is virtually to demand a bid within a certain time, to get a job—tree work, roofing paving—done so that another contractor can start his particular job on schedule. Or the patsy can dwell on the need to maintain the pace of building in order to obtain his next financing payment. Or, he may refer back to the referral system and the sub's reputation for giving bids without delays. When delays occur, the patsy naturally can phone the office to confront either the sub or his message taker: "Are you coming or not? You are slowing down all bids." And then he presses for a new time—that being as soon as possible. Usually the sub will be halfway out the door before the phone call is over if the patsy mixes in the strategy of displaying a bit of anger.

In using pressing strategies,* the patsy is ever mindful of keeping sufficient intervals between the visits to avoid meetings among subcontractors in the same bidding set. If the bidding is not too rushed, the patsy will use duration intervals as the best way of meeting elsewhereism demands on the sub. When there is only a day or so for bidding and the patsy's intervals are tight, he must use strategies like demanding that the sub leave work elsewhere and come over at a specific time. The contest between patsy and sub is most intense and inflexible at this moment. When a sub insists on coming in the morning, since that is when he starts a crew in the same area, the patsy (with an existing date) in the morning uses "white lies" and says he will not be on the site until the afternoon.

Another strategy is to ask how long the bidding will take, and if short enough, then try to squeeze the sub in around lunch hour, while shortening the interval between subs. If the squeeze is temporally too tight, meetings of subs can easily occur through a minor delay by any sub. Another time to squeeze in a bid may be after work. The patsy also may be forced to switch from a specific time on one day to the following morning, and then give the sub a duration time. Another strategy is to phone the sub if he does get to the site or ask if he is no longer tied up during the morning. This way if the other bidder doesn't show or his bidding is quick, the patsy can call in the sub who wants to come that same morning. The sub himself may agree to call in from the field to see if an opening, compatible with his freedom from elsewhere, occurs in the patsy's schedule. Similarly the patsy may give the sub two possible times and let the sub come then when he breaks free from other work.

When juggling subcontractors from the same bidding set, the patsy cannot foresee when elsewhereism will cause a delay resulting in a confrontation of subs on the site. The patsy can only trust that the subcontractor sticks to his promised time and comes neither early nor late. During the study this author experienced several close calls and one confrontation. The confrontation was due to one sub for doors coming a week late right within five minutes of another contractor. The man who came late felt he had a virtual monopoly in the county of door contracting and that allowed him vaguely to promise a time and then come when elsewhereism permitted. I looked at the late sub and excused myself by saying, "By golly, Pete, you were supposed to be here last week and this man this week." Pete answered, "Yeah, yeah, I know." But then Pete ignored the second sub. He started to disappear to the other side of the site when the other sub said, "Hi, Pete"—saying two more times before Pete returned a brief "hi" to his new competition. This exchange established their stratification in the door business. Pete was clearly not worried, as he later verified to me.

The author then normalized the situation by going through a painstaking detailing of the doors with the new contractor who was decidedly nervous while taking down the measurements, hardware and styles for eight doors. It is difficult to rush such complex detailing, but Pete stayed invisible during the necessary full hour. Later Pete said, "Oh, he doesn't worry me one bit, he's not our competition. I haven't seen him in a year. I didn't know who he was, I ran into him once a year ago. He doesn't worry me. He's not the guy I'm in competition with." Thus the first sub was continually discounted as important in the bidding system for Pete, who indicated the separateness of subs in a bid-set. Subs compete with people whom they do not know nor care to know about. And when their competition is forced on them, they tend to disavow the competition as a way of maintaining secretiveness regarding others in the system. (This tactic was born out in verbal confrontations when I mentioned one bidder to another in the same bidding set. Unless there was a clear attempt to break through the system, the sub would ignore whom I might mention as his competition.)

Subcontractors handle the patsy's pressing for a site visit in several ways when elsewhereism puts them on a tight schedule. Their standard strategy is to tell a little about demands of other bids or work, and then say they will try to squeeze the patsy in between these demands. "Well, I've got two this afternoon and two in the morning, so I'll come up in between." The squeeze may be explained by how the sub can detour from his schedule, without too much harm. These strategies may result in specific times for the site visit, but more usually result in vague times within day-long durations at a minimum. For example, a sequential strategy is to put the patsy on the list and see his job "toward the end of the week." Coupled with this may be the offer of a phone call just before the sub arrives—a strategy which helps to avoid confrontations with other subs. Vagueness does not allow a promised time which if unmet will dissatisfy a potential customer: it forces flexibility on an inflexible situation.

Big, busy contractors who use estimators have special tactics, warranted by having an edge over the patsy in power symmetry. Promising every day. "somebody" will come tomorrow is the typical way to try and keep the patsy happy. When somebody does not care, the excuse is that no one knows who was actually empowered to say he would. "They" put off the patsy til the "estimator returns from the field" and calls back, and then they switch the patsy from estimator to estimator forcing him to tell his details over and over, which withers his need to press for a visit. It is likely that yet another estimator finally shows. Big contractors do not court small, one house jobs, so they risk these displeasures. On the other hand, small subcontractors who cannot break away from commitments elsewhere, particularly ongoing jobs ("I'm pouring all day") may try to counter with low guesstimates. They hope the patsy will delay bidding on the chance of getting a low price.

Intense Pace vs. Little Elsewhereism: The contest for arranging time is less intense for both and quite flexible on the part of the subcontractor who is not under strong elsewhereism demands. This contractor meets the patsy at the latter's convenience as quickly as possible, and usually eagerly. He continually says, "At your convenience" to the patsy. This sub may operate under one of several conditions which free him from elsewhereism sufficiently to give fast bidding. He may simply be out of work or all his crews are not working. Watching crews may not take much time, and if he does work with them as the owner he is always somewhat free to bid any time. He may hire an estimator who is never really busy and therefore obligingly says, "whenever you want me." There may be two or three partners, so one is usually free to bid any time. The sub may be "rained out" with nothing to do but bid. These contractors unrelievedly "sweet talk" the patsy, not risking the slightest displeasure. It is this contractor rather than the patsy who may incessantly repeat the promised time of the site visit in order to ensure it as the patsy presses. If the patsy wishes to bid on Saturday, or after hours when elsewhereism is at a low, then even busy subs may sweet talk and meet the patsy at his convenience.

These subcontractors are eager to bid, so when the patsy presses for a bid, it is not unusual for the sub to say, "I can be there in ten minutes," or "right after lunch." This desire to give an "instant" bid reduces the patsy's pressuring to zero, but raises the problem of juggling the contractors to prevent their confrontation. Except for very popular subcontractors, bidding-sets tend to run the same on elsewhereism, if one is not busy none are and the chances are they all are free to come instantly. Thus, the patsy is faced with having to arrange for safer intervals between subs' site visits. But he need only cite an alternative time, a few hours or a half day away, since the agreeable sub immediately says, "Fine, at your convenience." In short, the patsy must quickly articulate his bidding-set so as not to be simultaneously inundated with subs.

The one condition that the patsy works under which prevents such spacing of subs is when an uninvited bidder just drops by. (I picked up many an overflow of bidding-sets from a construction across the street.) It is easy for subs to see that a construction job is about ready for their own specialty: "I was just up the street, so I thought I'd drop by to see if you wanted a bid too." Also, subs out of work drift by offering bids, or they solicit directly while going to work. The presence of uninvited subs results in many near-miss confrontations. They are hard to get rid of since, for the moment, elsewhereism makes little demand on them—especially the drifters, but even the busy uninvited subs usually give themselves ample time before returning from a bid.

Easy Pace vs. Elsewhereism: Whenever possible the patsy will try to bid far enough ahead in the pace of construction to take a scheduling of bids in stride, thus reducing the contest over arrangement of a site visit. Then, in facing the harried subcontractor, the patsy can be flexible "at his (the sub's) convenience,"—as within a week or so. In this situation, the sub uses various looseness strategies when arranging the site visit time. He'll "catch" the patsy's bid next time that he is in the area, if he lives at a distance or is working at the other end of the county. He'll phone next time he gets a chance to break away. He'll "catch" the bid toward the end of the week. Also, he is likely to misplace the patsy's request or message, or just forget his loose commitment since others are pressing. Without the patsy's pressing, he sounds vague "Let's see, today it's gonna be a little rough. Could we meet there . . . let's see . . . what's tomorrow, Thursday? Yeah, could we meet tomorrow morning? Would that be soon enough?"

This situation gives the sub a license to put off the patsy as long as possible while giving priority to more pressing jobs. If the patsy allows himself to be put off too long, he will find it necessary finally to start pressing for a site visit. He then finds himself under the first condition (discussed earlier) where pacing is competing with elsewhereism. To forestall this, the patsy may start pressing slightly before he is actually pressed so as not to be put off too long by a busy contractor. But he cannot press too much in advance because the sub easily can see that the construction is not yet ready for his bid.

Thus the patsy is forced by pacing to hold off pressing until the "best view" of current construction is ready for the bidding. The best view is when construction has reached the point where the sub can make the best measurements and see exactly where the job goes and how aspects of construction might interfere with it. Plumber, electrician and heating contractors are especially worried that after making a bid, new walls will spring up to create more work for them. To be certain that the site is really ready, subs may press the patsy for details to see if in their own judgment it actually is ready. For example, the roofer says, "Can I get up on the roof, I mean is there something to measure then?" To which the patsy has to reply, "We have about 4 more hours work," and that day he loses out to elsewhereism. Sometimes that pace picks up and covers the area for a job before the sub can get adequate measurement for his own job. Hence some subs will tell the patsy, who fortunately calls far in advance, "Don't go too far before we take measurements, so you do not have to rip stuff out." (This quote is from a masonry contractor who was worried about the framing around a fireplace. He gave the patsy exact measurements for framing as part of his bid.) Sometimes the pace is a toss-up: for example, one foundation contractor wanted to go over the site before it was dozed but another contractor wanted to see it after it was dozed. Bidding far in advance solves these problems and allows the sub temporally to adjust his elsewhereism to the pace of the patsy's construction. The time of the site visit is not contested, but is flexibly arranged.

Easy Pace and Little Elsewhereism: This condition happens seldom as there is little stimulus for the patsy to call for bids when he is not ready and the subcontractors can come at almost any time. The few times that it occurs are when it is raining (everyone is rained out), or a bidding-set is in a slack period, or a type of subcontractor is being edged out of business by lack of demand for his product (a cabinet maker facing the prefabrication market). The patsy will usually wait until the pace picks up, realizing that the delays will not be caused by elsewhereism—subs will rush over at his convenience. Under the first condition (rain), the patsy may ask for bids and the subs acquiesce just to keep busy.

One other condition forces bidding far ahead of time for the patsy: financing. He may have to get some bids before he is granted a construction loan or have some wiring in the walls before he can get a second payment (or at least show the job has been contracted for).

Can't Say No

In discussing the fitting process that leads up to a site visit, several aspects of the job were noted which might lead a subcontractor legitimately to say "no" to giving a bid. Very few contractors said "yes" to a visit when there was not fit, and these desperately needed work. But when a job appears to fit (on the dimensions of type of work, details, price and location) it is hard for a subcontractor to say "no." Yet he may wish to for private reasons—some jobs he just does not like. Or more likely he holds off for elsewhereism reasons—he is holding off for more lucrative jobs or is too busy on other type jobs.

But the basic reason a subcontractor has a hard time saying "No" to the site visit is because of the referral system. He always wishes to appear willing to do a job under any conditions—to maintain an image of providing service. With this image he maintains his place in the referral system. He ensures future callbacks from the patsy and his friends, and referrals from other subs. Should the word go out that he is not a willing worker or avoids certain jobs that appeared to fit, then he will be dropped in others' referrals. Once dropped he does not have a stockpile of potential jobs from which to select for increased profit and to keep the morale of his crew up, and cannot keep a crew working during slack seasons when any job sounds good. Thus, to say "no" may close out much future business.

In the beginning particularly, this propensity poses a problem for the patsy, for then he may depend on subs who later will say "no" by their action while still talking "yes." Thus the construction will be delayed. At minimum, the time of all concerned is wasted. These delays occur in three ways: the subcontractor does not call back after promising during the patsy's initial call; he does not show up for the site visit after arranging for it; and on the site he develops a bid which indicates "no."

No Callback: The promise for a callback may be made by the message taker or the contractor. In both cases the reason is almost always elsewhereism. The message taker will say, "He was around, but I don't see him, may I have him call you back?" or "Can he call you back about 4 this afternoon?" The sub himself will say something to the effect that he is busy, or rushing out, or would like to look at the job and will call back for details. Regardless of who takes the message, the subcontractor never calls back. If the patsy calls again, he receives the same affable promises but no return call. Thus, hardwood floor contractors are very hard to get callbacks from: they have so much work stockpiled that they only call back the most preferred type of job when the current job is almost finished.

On Site "no": If the sub reaches the point of a site visit—either because he does not know how to act "no" or has not yet decided "no"—he may apply several standard strategies to get himself off the hook without a blunt refusal. One is to explain how the patsy really could do it himself and save money. (A formica contractor and painter both did this to avoid specialty jobs that entailed too much work and liability for mistakes. They were set up only to whizz through standard jobs.) The sub also can simply act "no" by showing little or no interest, a cursory approach to details.

His lack of "grab" generates a feeling of distrust in the patsy. The sub also can bid himself out of the competition by going too low or too high. The low bid makes his work look too cheap to trust on quality and time; the high bid virtually assures its refusal. If the high bid is taken, the sub makes extra money for his trouble. These strategies may be followed with a recommendation that the patsy use the bidding system for comparisons—the sub knowing they will not then be chosen.

Not show up: The most frequent way of acting "no" is to agree to a site visit at the convenience of the patsy, but not show up. This, of course, costs the patsy much time. At this point—the waiting having perhaps delayed aspects of construction—the patsy realizes in anger that the sub had got rid of him "pleasantly." At this point the patsy can just accept the fact that he has lost out in the competition with other people—particularly general contractors—elsewhere or he can still press the sub for a site visit by calling again. Invariably he will get a polite apology and an excuse. Most excuses will point to demands elsewhere which are of more priority than the patsy's job.* "The X is out, I'll leave a message." "Oh, I've been so busy, I'd forgotten it." Or, from an estimator, "I have five more calls to make and then I'll try and catch you." "I am two weeks behind in my calls and my work." "I got stuck up in the country." "We had an emergency." Often subs say they still are coming and even set up a new time, "I'll make it for sure this morning; boy I have been busy," but never show up. Once a sub had decided to work on jobs elsewhere, he seldom changes his mind until work runs out. Thus setting up a new time usually means only further delay and loss of time for the patsy.

Given a modicum of intelligence, the patsy soon learns when he cannot compete with elsewhereism and that minimal competitiveness is necessary in order to establish a sufficient power symmetry to have some future control over the job. Unless there is this power symmetry, the sub will probably prove unreliable at a moment's notice—as he races elsewhere and drops his commitments to the patsy. The patsy learns not to press too hard. After receiving a fair number of elsewhereism excuses, it no longer becomes important for him to inquire why a sub did not show up. The patsy learns that the sub with whom he can work is among those who do show up, and he proceeds to work with those subs who do. "No show" becomes an accepted part of the bidding process as he learns that contractors often avoid saying "no."

Now in the next chapter, we turn to what happens when the subcontractor actually turns up for the site visit and details the job for a bid.

III

Detailing the Job

The site visit is made for the purpose of a full detailing of the job. The temporal focus of detailing is on making agreeable aspects of the job—particularly quality—to both patsy and subcontractor so that accurate estimates of time and cost can be made for calculating a bid. As the detailing proceeds, both patsy and subcontractor are trying to achieve several ends.

The subcontractor is trying to gain trust and acceptance on his type of job from the patsy—and his job may include some redesigning of the patsy's job. The sub may even be trying to find ways to *beat the bidding system* through neutralizing competition. Both sub and patsy are trying to detail as clearly as possible the work involved in the job, so that each will know what work the other is responsible for—just what carpentry the patsy is to do in order to prepare for the sub's job, and just what the sub is to do. If the carpentry is not ready, a sub cannot proceed since he is not allowed to do carpentry unless it directly relates to his job. Also, while a plumber can put in a block to hold a pipe, a heating contractor cannot make frames for heating vents. Even if the sub is allowed to do the work which the patsy has agreed to do, if he did so, this would change the time and costs of the job.

On his side the patsy is doing more than detailing specifications. He is resisting both being hooked and having his plans redesigned beyond his liking. As the judge of a bidding set, he tries to keep bids comparable on details of the job—particularly with respect to quality, time and costs. For simple bids like a fireplace, it takes little time and effort to bring bids into comparability on all details. For complex bids, the patsy is trying to gain *information* from the sub on how the job is done; trying to judge the *risk* in the job and *trust* in the subcontractor in using each contractor; and trying to keep the subcontractors, whose talk drift to personal experiences, *focused* on the detailing. (These personal breakthroughs in on the job conversations during work will be discussed in the chapter on generalling.)

Trust, which both the patsy and the subcontractor are trying to build up, is the crucial aspect of the site visit. While other aspects such as detailing specifications and job, and negotiating time and price, have much value in their own right; they also are the source of the patsy's building up to sufficient degrees of trust in each sub. As each aspect of detailing is resolved in some manner, its resolution becomes a part of the trust which the patsy can place in the sub. When resolving these matters, the sub in turn *may* be trying to engender trust by using some form of interactional strategy. However, mainly he does not, but is just himself which has consequences for the patsy's trust in him. (Thus we shall mainly be discussing the consequences of subcontractors' styles in bidding rather than subcontractors as manipulators.)

Trust is crucial because what the chosen subcontractor is doing for the patsy is creating visual transformation in the patsy's planned construction. This transformation is truly fantastic: there is nothing except imagery, and when done the building looks like it has always been there! Then the patsy must live with it for a long time. Therefore, the patsy's aim is to make sure that the transformation will be close to the quality and taste he wishes for his home. After it is built it will reflect the patsy's taste, and no one will realize that he did not wish it so. To change it then requires another transformation with the same detailing properties of subcontracting; that is, correcting a job can be like ordering a new one.

Now let us consider these aspects—detailing specifications, time and cost—of a site visit with regard to their value for developing trust in a bid. At the time of the site visit, elsewhereism is at a minimum and power symmetry is balanced in favor of the patsy. The patsy is in complete command on his own terrain, exercising the power of choice and judgment over the sub's attempt to land a job. The sub is cut off from the demands of work and others elsewhere, unless they know where to reach him by phone. (This occurred in less than 10% of the visits to the author's site.) The patsy, therefore, has complete claim on the sub's attention to detailing the job. This is likely to be the only time he will ever have such a claim! (The few phone calls that breached this solid attention, the author cut short.)

Detailing Specifications

In detailing specifications for the job the patsy is engaged in developing a *mutual understanding* with the sub of what the job entails, *warding off the redesigning* attempts of the sub, *information gaining* from the sub, *judging and juggling risks* and *evaluating the sub's attitude*. As these aspects of detailing simultaneously proceed, as we have said, they closely relate to each other and add or detract to a total feeling of trust in the sub.

Understanding the Job: Some jobs are relatively complex (cabinets, tiles, doors, heating, plumbing and electricity) and some are fairly simple (roofing, masonry, glazing, drywall and hardwood floors). Keeping this variable condition in mind, the patsy gauges how fast and how systematically the subcontractor sizes up the job with a grasp and understanding that rings true to the patsy. Thus if he sizes up the job fast, but his rendition of it is off and/or unsystematic, he is seen as potentially a maker of mistakes—perhaps even sloppy and not to be fully trusted, especially regarding quality. But if he sizes up fast, especially a complex job, with perceptive understanding of what the patsy wants, he is likely to win trust immediately. If he couples this with "talking quality" at a decent price, then he is in solid; only grossly negative features can shake the trust engendered on first impression.

"Being systematic" is accomplished in various ways. The sub may take careful notes, as he talks through the job from beginning to end with the patsy; then he reads back the notes to verify the details while explaining all necessary points. He does this expeditiously and listens to the patsy carefully when he is not talking so as not to miss a word of the patsy's intent. During moments of silence he appears to be thinking through the job or problem that has arisen, and then offers sensible solutions in a manner that suggests further discussion with the patsy. The latter thereby develops the feeling that he can work through the problem with the sub and trusts him because of his effort.

The subcontractor also wins trust by attending immediately and efficiently to detailing the job as opposed to detracting by digressions, about what he is doing elsewhere or about recent interesting experiences. Only after the detailing does a personal digression of the work situation seem warranted and usually it is linked with a friendly goodbye and promised time for submitting the bid. Otherwise, when the sub floods out with personal experiences, the patsy is forced to refocus the sub back on detailing. Such an interruption raises questions as to how well the sub will stick with getting the job done. Personal digressions can be exasperating, burdensome, cause annoying delays to the job and indicate unsustained grasp of the patsy's situation. They only make sense at the end of the business talk, acting to reduce interactional tension.

The subcontractor's understanding of the temporal articulation of his job with others' jobs is vital to the patsy's trust in him. It is not only a matter of where his job fits into a succession of jobs so as not to cause unfortunate delays, but where the details of his job articulate with the details of other subcontractors' jobs. Whether subcontractors will directly articulate jobs among themselves or under the patsy's direction (such as the excavation contractor digging partly for the foundation contractor, or the tile contractor working with the plumber understandingly and willingly) is crucial to entrusting the job to them.

The contractor who arrives for the site visit early or late and thereby misses the patsy for the joint detailing of the job, has two choices (other than to wait if he's early). He can try to detail on his own if the job is simple enough or apparent enough or he can leave and return. Either way he undermines trust in his temporal reliability, and possibly so in the quality of his work. Without the patsy, simple detailing can be done, such as roofing or paving. The sub measures and calls later and explains his understanding clearly over the phone. Short of this condition, the patsy is likely not to hear about the job he wants, but rather about the sub's preconceptions of the ordinary standard job. Hence the patsy will not trust him to comply with the special details of this particular job.

Risk: Closely linked with understanding the job is the detailing out of the degree of risk—as bargained for on diverse aspects of the job to be accomplished. Quality is, of course, intimately bound to this. Quality entails value expectations, with some degree of risk embodied in the work expectations. To what degree there is risk in obtaining the desired quality is hard for the patsy to calculate, since at the extremes a sub may simply do a poor or first-rate job regardless of work conditions. But the kinds of materials he will use, the time and the cost (price) he allows for the job all indicate whether he allows himself "space" to do a decent job. Subcontractors often will offer to show work done on other houses, so the patsy can judge their craftsmanship. Also the patsy often uses "vouched for" referrals to help gauge his risk.

Other, more concrete risks can be forstalled during the site visit. The patsy can cover himself by assuring that details are straight, that there can or will be no surprise extras, that code requirements will be met and that a contract will hold.

For all jobs, there can exist a communications gap between what the patsy wants and what the sub understands and intends to give. Closing this gap reduces the risk in getting an abortive or wrong job; it also increases the patsy's trust in the sub. To close the gap may be difficult as some subs just never seize the details for specialized jobs—particularly when they are so used to doing a standard job—such as special dry-wall work abutting redwood, or special tile work in the kitchen counters or the bathroom, or special masonry walls. These standard job-oriented subs represent risk. For complex jobs, the patsy may discover that the sub should submit a detailed contract to reduce all chances of wrong or missed details. One cabinet maker went over, for verifications by the patsy, the complex instructions that he was going to submit to his shop foreman. In another case, involving a simple job, the patsy made a roofer measure the roof deck twice to be sure he planned for enough materials, because the sub seemed sloppy in his first measurement. Sometimes it is obvious that gaps in the details can easily be closed when the job starts, so there is no risk in each party somewhat talking past the other.

If there is a risk that the gap in detailing will result in "extras" being charged, then it will pay one or the other to review the details several times. One electrical contractor—for a complex bid—went through the details several times to reduce the risk of adding needed extras which may so easily enter into an electrical job: as the job unfolds, new possibilities can open up for outlets, switches, etc., and he was worried about accusations of raising his bid with extras. The result was, again, an increment in the patsy's trust. Similarly, the cabinet maker wants to be sure who will hang the cabinets, since it is typically his job, not the carpenter's. One dry-wall contractor carefully detailed the types of nails to be used: the rinkshank nail would add more costs to the job but would allay all danger of nails later popping out of the wall. Thus he forestalled adding this type of nail as an extra later and won trust. These efforts at avoiding extras win trust, because a well-known standard strategy for getting a bid chosen is to bid low then make up the profit through emergent, "enforced" extras for achieving minimum quality during the actual job.

It is also important to detail the responsibility for risks of the job for third parties who may later become involved. One dimension is the liability for repairs during the guarantee period for new owners. In setting forth details contractors may require certain aspects of a job as he and the patsy shift risk responsibilities back and forth between each other. One paver insisted on a cement driveway since liability to the homeowner lasts seven years. The roofers insisted on a minimum number of layers of tar paper to get themselves through the guarantee period. (At this point, the patsy must be sure he is dealing with a licensed contractor, otherwise he will be responsible for everything.) In another case the excavation contractor would not formally accept the grade ratio for creating rock wall, although he felt the chance of it crumbing was nil. But the patsy accepted it on his informal advice to get the work going.

Another dimension is the passing of building code requirements. The electrician wanted a certain spacing and number of outlets to ensure that the job would be passed by the building inspector. Again, the plumber and mason stated that it was the patsy's risk to obtain inspection approval if they proceeded as he wished. The prudent patsy complies with such counsel in order to shift the whole risk of meeting codes to their contractors' shoulders and so that adjustments will be at their expense. One surveyor did a simple contour map, saying it was enough and taking the risk that it would pass inspection. He saved the patsy money. In the case of the septic tank, part of the bid was for the sewer contractor to obtain the permit saying, in effect, his job was acceptable to health codes. Once this aspect of risk is settled, trust in the contractor follows. It is possible also to get a feasibility detailing from a sub, to judge risk of passing codes and who can best take the risk. Once these crucial risks of Liability and communication are settled—these may include detailed contracts with written guarantees—trust in the contractor builds up if his understanding appeared competent.

Complex bids (doors, plumbing, electricity, cabinets, heating) usually will be given in the form of a precisely detailed contract requiring both signatures. Simple bids (glass roofs, foundations, drywalls, painting, paving jobs) come in a verbal form or with a price scribbled on the back of a business card. The latter form of bid entails the most risk in forcing performance and thereby the most trust in the sub's reliability. But this reliability is seldom an issue if trust in the sub has already begun to develop, otherwise the risk of a card or verbal bid being taken as a contract seems besides the point.

Redesigning Process: When detailing the understanding and risks of a job, a subtle process of redesigning often occurs. This can undermine trust and can edge the balance of power symmetry in favor of the subcontractor. The redesigning process occurs when the subcontractor, by going over details and making suggestions, actually and significantly redesigns the job: that is, the job will be significantly different than the patsy intended. This may happen imperceptibly. The patsy may like the changes or may find quite annoying his loss of control over the job.

If the latter, he may privately disqualify the sub as someone who changes the details and is not to be trusted to stick to the job as detailed. (In my own case the latter was the most likely reaction when the suggestions and changes were those of either a taste nature or technical suggestions having taste implications.)

Technical and taste suggestions are basic ingredients of the redesigning process. Technical suggestions flow normally with the discussion during detailing. They build up to redesigning under the following conditions. They are part of either the job requirements or the code requirements for which the patsy did not plan. Or, the patsy has the imagery for the job but must ask advice on the technical details, and this information feeds back to change his imagery. Taste suggestions arise, although infrequently when the subcontractor cannot resist his own design ideas—albeit they are usually standard ones.

Job requirements vary according to whether the subcontractor is system or structure-oriented. A system-oriented subcontractor is a man who puts in an electrical, heating or plumbing system into the house. These systems have three variables—input, output and vent—and each must be done in a relationship to the other, as required by the building code. If the house is designed so that the system subs cannot fulfill system requirements they will simply ask the patsy to redesign a portion of the house. They will ask to make a wall inches wider, to move a wall, or redesign a wall so as to put a vent in a different place or make electrical runs easier, to put the heating outlet in a different place, or to put the cold air return in a different place. And they will ask these things of the patsy assuming that their systems have priority over his structure, which will have to be changed to comply with their system so that the system will run properly and pass inspection. Very seldom can they bend the systems to fit his structure. So the patsy often will be pressured to redesign some things for technical reasons. However, when the subs' requests appear to be excessive and unnecessary, the patsy can counter with "Try and think of another way to do it."

A structure-oriented subcontractor is a man who works with a structure: he builds a chimney, puts on a roof, puts in a driveway, puts up sheetrock or lays a foundation. These are simple static structures. Now, these men often are very rule-oriented in the sense that they are guided by standard patterns of resolving their jobs. This allows them to work very fast. When the patsy wants one of these men to do something special, often they will tend immediately to cite the rule for how a chimney is made, sheetrock is hung or a roof or foundation is laid.

Then an insistent patsy must tell them that they are going to bend the rules a little bit—to see if they can "take it."

The patsy finds it is far easier to resist the normative pressure of a structure-oriented contractor who wishes to redesign the job. Sometimes the patsy likes a suggestion or realizes its necessity, but usually when the sub tries to standardize a special job, it eliminates its specialized stylistic features. (For example, a door contractor rebelled against cutting all doors two inches below their standard height so that they would be on the same level as window heights to maintain—the patsy thought—a pleasing level line to the eye.) Then the contractor can be told that this is how he must do it and that the patsy does realize that special jobs may cost more: the sub then decides whether or not he will submit a bid. (For example, the roofer would not do special skylights for the author. He believed only standard ones were leakproof.)

Both system and structural-oriented subcontractors suggest technical changes, but only the latter tend to make taste suggestions. Usually they do this because they cannot resist doing so. The few who cannot resist are far outnumbered by the many who steer clear of all taste discussion for fear of offending or annoying the patsy, hence losing the bid.

Taste suggestions vary from discrete aspects to a total view of the job. The discrete suggestions usually are the standard tastes "going around." (The subs dealing with tile, floor, formica and sheet rock alt couldn't resist stating the latest everyday tastes when it came to choosing colors, textures and sizes. The glazing contractor pushed aluminum instead of wood sash windows.) Besides being routine, their tastes typically are absent of relationship to their job or to the total project. However, some contractors wish to see their work in its total context their way. These are the frustrated home designers,

(In particular, one masonry contractor helping with a special firebox and used brick wall kept trying to impose his taste on the total picture. The author kept saying "no" to his "I think you ought to do it this way, if you really want it to look good." Finally to put him in his place I said, "That would really look terrible." But as we went over the job again, the sub tried to redesign the work, using as leverage the core requirements that he had to meet for inspection. Finally, after a compromise between our tastes, and risk on the inspection code, we agreed on the details; to which the sub said, "Well, I guess it's just like women, each man has his own taste." The building inspector, another frustrated architect, tried to use the code to force me into using some of his tasteful designs for porches and doors. He capped his approach with the notion I would also save money.)

In backing up their taste with building code requirements, these people try to make taste technical; hence in their view they legitimate the style. (In my experience, they were not clever enough to keep technical requirements separate from taste; it was all one.) In the patsy's view, the redesigning can be mediocre, even if more efficient and less costly. To resist the sub's suggestions is not difficult once the patsy brings his style up to code and/or job requirements. But working out a compromise is not easy, since when the sub's taste is ignored he is likely to become very rigid on requirements that otherwise he would give way on, (At one point the building inspector changed his rendition of the code to fit his own porch plan for my house.)

In the flow of technical suggestions to the patsy on jobs for which he has imagery but not sufficient details, the contractor's usual stance is to sound hesitant when making suggestions, and to withdraw them immediately if it appears the patsy does not like them—and then cover with more sweet talk. In providing information to put the job together, the subs have no illusion either about their positions and the patsy's prerogatives, and they do not wish to lose the bid. (Power symmetry at this moment is in the patsy's favor.) Most subs will try to implement just about anything the patsy wants, as long as he pays for it. Thus, in providing alternative details, they try to embody the patsy's imagery as fully as possible, even as they may happen to be redesigning the job at points where his imagery was too naive to work. (For example, I knew the style of cabinets to be made, but not the details and dimensions to get in all the various drawers, storage and appliances. One cabinet maker was an expert in tailoring to my imagery, as he redesigned our dimensions and the total layout. A tile contractor made my planned imagery better by appraising me of special tiles I did not know of, and redesigning the job to use them.) This kind of sensible redesigning engenders trust, since the patsy's imagery is enhanced by details happily agreed upon.

Some contractors, so chary of offending the patsy, provide a series of logical options, many of which obviously are not workable. The patsy then can by comparative perspective choose among those options that would work, and appears then to be designing the job himself. At this point the sub asks what the patsy wants and then agrees with his choice. (My excavation contractor and formica contractor used this technique well, since their jobs were so simple as to provide many options.

This strategy also worked for subs who had little ability to embody imagery with details. They provided an array of details which I could use to formulate my imagery.)

Very few contractors will use a request for unknown details as a way of sowing doubt as to the feasibility of the patsy's imagery and thereby gain the power to redesign the job. But when this happens, the patsy is quite vulnerable to being directed by the sub's advice. The patsy is not quite sure of what he is doing—as opposed to the conditions mentioned above when he has sufficient knowledge of the details of a job—and he is easily dissuaded from his wishes and taste. If he can resist this expertise long enough (it is difficult), his best strategy for warding off the expert is to ask for alternatives. The expert can be put on the spot by intimations that he is not very expert if he cannot muster several alternatives. It is under this condition of the patsy's having an imagery (a goal) but not knowing the details (means) that he loses his edge in power symmetry and that changes most likely will be beyond his wishes. This is the classic expert-layman predicament that is so pervasive in all such relationships, especially professional ones, in which the patsy is vulnerable to being led (perhaps, even, exploited) by the expert.

Information Gaining: Besides the redesigning process, when asking for technical details on how to make imagery actual, another basic process of the expert-layman relationship is emergent. This is the information gaining process. Learning about the mix between technical and taste details and about the various alternatives which exist, the patsy also learns about the general kind of job rather than only about his in particular. Also, many subcontractors typically play teacher, giving suggestions to the patsy. The patsy soon learns which subs like to teach; therefore, how much each can be pumped for information.

The patsy usually wishes to gain information for two to three reasons besides the detailing and designing of the job's imagery. He wishes to gain administrative and inspection control over the job, which may either give him the edge in power symmetry or prevent his losing what edge he has. Also he may wish to learn how to do-it-himself.

By *administrative control* is meant that he learns how the job is paced and its succession of activities; this gives him a measure of temporal control. He learns enough to administer the sub: the patsy can spot unrealistic and realistic delays, hurried vs. well-paced work that is so important for quality, and in objecting to the former can make cogent suggestions about how the pace can be maintained. It is hard for a sub to refuse such cogent suggestions; in effect either he is controlled or his control is reduced. The patsy can also better temporally articulate a particular job, such as plumbing, with the other ongoing jobs of the total construction. In doing so he can again cogently repace the sub's plans for his job. This latter source of temporal control is vital in generalling the job and its relation to the total construction, as we shall see in the Chapter VI. Some subs will resent and resist such temporal control but most take it good-naturedly, bow to its cogency and acknowledge it as one less barrier to getting their job temporally well done, hence to insure a happy patsy—and *get paid!*

By *inspection control* is meant that the patsy learns enough about what will go on so that he can inspect the work that is being done. This gives him quality control. In learning the processes of a job, the patsy learns how to spot subs who will cheat on or short-cut the job. Some subs will even show how cheating occurs so that the patsy can watch other bidders, (For example, one roofer explained how a 3 ply roof can actually end up 2-1/2 ply by the way it is laid.)

After the patsy learns what is correct for the job, it is hard for the sub in bidding on work not to yield to his temporal and quality controls. It is also hard for the sub to delay or change details that deter from quality, since the patsy knows what is happening and can counter the sub's attempts at temporal and quality departure.

Subs realize this, and resist giving information beyond whatever details are absolutely necessary to reduce the patsy's control over them. As the patsy pumps, they hedge. Some subs, however, like to provide a patsy with enough information to exert inspection controls. They wish to do a good job and appreciate the patsy who, armed with information, is capable of understanding and evaluating a quality job. Since their jobs will be quality jobs, they regard good evaluations with pride.

Subs know that if they give too much information—and especially if they give a design—they are teaching the patsy to do the job himself. In essence, the patsy could do any of the sub's less skilled jobs if he wished to take the time and/or could find a crew to help and could rent the necessary tools. All he needs is a little know-how from experienced people. (For example, subs trained me to do a formica job, build cabinets, put up wallboard, build a swimming pool and so forth,) Subs will vary on how guarded they are with the necessary details on how to do the job and where to rent the tools. The sub who likes to teach cannot resist the patsy who is a willing student. Others will not even give enough in orientation to make a bid clear, no matter how much they are pumped. They emanate a 'trust my expertise' message. Most subs fall in between. The same understanding and implementation of imagery that they give—which engenders trust—often contains enough information for the beginning stages of a do-it-yourself job. The patsy can then, if he wishes, probe to round out the details. At this point, seemingly simple jobs may actually become so skilled or complex that the patsy clearly would not risk doing it himself and so jeopardize quality. Realizing this subs may make the job appear too complex as a scare tactic to do-it-yourself patsies. (One sub who wanted a paving job very much said, after complexly explaining exactly how he did it, "Now you can do it yourself," to which I replied, "No, I just want to be able to watch what is done.")

Every now and then a sub definitely will not wish the job—because of elsewhereism or because the job would not pay—and try to get out of it by teaching the patsy to do-it-himself, at each step telling him how easy the job is. Thus, he does not harm referrals to himself and lose future callbacks. The sub will even pump the patsy about the job, to make sure he will cover all details of how to do it. (One electrician did this for a small extra job, after he finished the big one—hanging the fixtures. A formica contractor also did this to avoid a special job.) Sometimes a sub will try to tell the patsy how to do another type of sub's job, in order to get the bid for his own job by engendering trust in the patsy. (For example, the dry wall contractor explained the best way to paint his drywalls, and the door contractor was an expert in telling how to trim floor moldings oneself.)

In my case, information gaining was mainly directed at administration and inspection control, beside the technical detailing of imagery. There are, however, two other consequences of information gaining that can be used purposefully, although most often they are subtle by-products of the process. A sub who knows a lot about either his work or other kinds of jobs can be pumped for information concerning the feasibility of a job, that has yet to be planned but is growing in contemplation. (For example, an excavator for the foundations walked around the corner to see if a new lot could possibly hold a septic tank system, which he subsequently did. Or, the roofer judged the feasibility of water-proofing the back of the house with a layer of hot tarred paper under the siding.)

The second consequence is that when a sub is flooding out with details and ideas for his job, he can give the patsy ideas for new jobs for the house and even give bids on the uncontemplated job if the realization strikes him fast enough. (For example, the tile contractor in detailing the bath happened to mention in passing how good kitchen counters looked in the same tiles that I had chosen. His ideas sounded so convincing that I took a bid on the counters too. At that point the conversation strayed to tiling the kitchen floor, which led to another bid; after this he said that he did a lot of entry ways in the same terracotta—yet another bid for a great idea. The glazing contractor happened to tell of the fancy glass that he was currently using with terracota tiles, so I took another bid for windows in the entryway.)

The information gaining process is actually consultative, but the consultant is not paid. Since many subcontractors advertise free estimates and this service is a condition of the normal bidding system, virtually none can expect to receive the payment. As one sub said to questioning about how he could give out so much free advice, "Oh, no, you can't consult. They'll just say it's a free estimate and won't pay you." The subs can only be careful that they are going out to detail bonafide jobs, since the wise patsy can easily use this consultative system to learn how to do jobs himself around the house. (Checking out the patsy as noted earlier, is largely accomplished by the referral system.) The sub's only other alternative is to consult for many but not do jobs, therefore not bid, and to make this clear to his client. This rarely occurs, because other subs will consult for nothing in order to bid.

The information gaining process becomes structured according to the number of bids taken on a job. For my study, it was structured by the normal temporal pacing of bidding of subcontractors, as opposed to bidding for quick jobs. I developed a three stage process, a stage for each bid. These stages provided for an accumulation and comparison of information. Information gaining varies by the necessity of what is needed at each stage to detail the job as the patsy progresses from first to third bid. The patsy's relationship with the sub also varies for each stage; he achieves more favorable power symmetry as the bidding progresses.

With the first subcontractor, the patsy plays student. If he knows about a sub who likes to teach he asks him for the first bid because he is in great need of information. Sometimes the initial call will indicate how much a sub will teach, and then he can be asked on the site first while discussing temporal arrangements. (For the heating bid, I first asked a sub who loved to talk and to design technically with no pretensions about taste.) Usually, however, the patsy must take his chances on whether the sub will gladly teach versus how much he will have to be forcefully pumped for information.

The teaching is started off by the patsy's request for advice and specifications concerning the implementing of imagery. This is, by and large, a manifest request for learning, the diverse consequences (noted above) remaining surreptitious. Insofar as the sub comes through with good alternatives and details on procedures, materials, and risks indicating his understanding, he gains the patsy's trust. (I said after the first plumber, "He won our trust. He explained aspects of the job to us carefully, answered all our questions so we understood how it was to be done and we could make decisions.") In playing student, of course, the patsy loses the edge in power symmetry for the time being through his clear deference to the first sub's expertise. Inflated by his teaching, the sub may pre-empt decisions; but the patsy only loses the edge for a short period, since if the first sub actually is chosen then the patsy comes back to the relationship as a well trained student with knowledge of what he wants done and with powers of administration and inspection.

With the second subcontractor, the patsy now immediately engages in a relationship involving give-and-take discussion about details of the job. By now the patsy has an idea of what he wants and some categories and meanings for the details. He tells the sub and asks him for comparative ideas on what he would do. In essence, the patsy asks for possible counter advice in an effort to check and round out his comparative knowledge of the job. For example, he tells the sub what BTU will be used for the air furnace and asks for his agreement, opinion and suggestions. Thus, the patsy's need for information has lessened and his ability to state what he wants has grown so that he loses no edge in power symmetry; he is not vulnerable, but in more or even complete control. He is more certain about what degree of trust to place in the sub, being now backed up by his accumulation of knowledge.

With the third contractor, the patsy now knows almost, if not exactly what he wants, and simply makes the details known to the contractor. He plays wise with the implicit idea that since no discussion is required, any initiated on the part of the subcontractor. Thus, new information gained is not sought yet is listened to. Usually the subs will offer alternatives already known, to which the patsy responds "I have considered these," and firmly holds onto his edge of power and the image that he wants and means to get through control of the job. Subs often like to be told exactly what to bid on, it saves them time; and when they just bid with understanding at this stage, they win the patsy's trust. However, if they challenge with new information, trust is not lost; but if their alternative is already known, they may unawarily undermine trust through implicit redesigning with already considered and discounted details.

Playing wise at this stage of the information gaining process can result in an information dividend! Contractors tend to tell the patsy "in-the-know" of the more subtle aspects of the job which go beyond the necessary information for timing the designing, details and quality. For example, subtle differences in work can be related that will not affect the job for years to come: such as that the difference between thin and thick coats of hot tar on a roof means a few additional years of resistance to the sun, or the best angles of clean-outs on the plumbing waste lines or the best under rock material for driveways. This information as passed between two "wise men" can feed back into the job details to reduce trouble in later years and/or helps the patsy to watch other contractors who may cut comers in detailing.

But playing wise can be dangerous for the patsy at the second, and even the third state of the information gaining process, until he is clearly "in-the-know." If he plays wise prematurely, implicit in this stance is that he is aware of the basic detailing assumptions which are built into the job. If he is not aware, he may discover these "obvious" (to those in-the-know) matters later when contemplating the finished job. Displeasing aspects cannot then be easily removed. If he questions the sub about these unexpected details, the sub will probably register surprise that the patsy did not know the details. The sub thought he could skip them. Examples are the basic thickness dimensions of sheet rock, the volume capacity of the pipe sizes, the standard door jam, the standard way cabinet doors are hinged and the typical heights of fireplace hearths.

To the degree that the patsy gets through this information gaining process for a particular job, he develops confidence in his ability to choose and to trust a sub for the job. His confidence is based on his knowledge of how to administer and to inspect the coming job, and the edge that this gives him in power symmetry.

Attitudes: As the patsy sifts through each subcontractor's understanding and efforts at redesigning, the information he has gained and the risks of the job, the trust he develops in each sub is toned by the attitudes of the subcontractor. Immediately apparent will be how the subcontractor responds to and handles the tension of serious detailing. Insofar as the sub can joke or be jovial while keeping focused on the serious detailing, he reduces tension without losing the focus of work. This attitude enhances trust. It serves as an indicator of how the sub will handle the tensions of the forthcoming job. His effort will be to keep the job going while keeping the surrounding social climate compatible.

Another attitude of concern is how the sub will handle adverse difficulties during the job. As the patsy details, sticky unanswerable points may arise which could slow or stop the job and result in extra costs. Many subs will have the attitude, "Oh, I'm not scared," or, "the impossible takes a little longer, that's all." They enhance trust in essence by emitting assurance of their ability to handle and absorb difficulty within the confines of the existing bid. The sub who does not reassure rather immediately, and seems to hesitate, undermines trust concerning his sticking to the contract.

As we have said, in the initial call most subs will start off with standard "sweet talking." Then on the site visit, the sub's actions show to what degree this talk reflects sincere attitudes, and not merely a come-on and softening up action. The sub can support his talk with enthusiasm for the job, and agreeableness to what the patsy wants done, perhaps also with a friendly comment on the taste of the house, and a willingness to begin on time at the patsy's orders so as not to interfere with other jobs. These actions win trust. They acknowledge the nature of the job and the patsy's control over it. Not to make them does not necessarily detract from the trust enhanced by a particular sub on other dimensions, but he may lose out in the competition which is based on these multiple sources of trust. "It is just better to work with compatible people," said several patsies, "it takes the pressure off."

An example of the joviality that puts both patsy and sub at ease was one plumber's response to the patsy's desire to put colored (not white) fixtures in the bathroom. "Oh, that's good, because you have to have color somewhere. If you don't, CORE will discover it and there'll be pickets all over the place." The ingenuous sub who emits these attitudes too sweetly, of course, runs the risk of being suspect and of losing trust.

In contrast to extreme sweet talking, among all the bidders only one sub got angry and sacrificed me, his customer, to it. His head shaking with anger, he furiously said regarding the masonry job: "We get too many of these dirty jobs. I want to do everything or nothing. I'm not going to do anymore of this dirty work." He handed me his bid and huffed off.

Detailing Time

Although the detailing of specifications is concluded, questions remain as to when the sub can do the job. The major ones are whether the sub's promised timing fits the requirements of the construction and whether he can be trusted to comply with the promised time—a very crucial aspect generally for the total job.

Fitting the job to the temporal requirements of construction is accomplished through negotiation both of its initiation and duration. This negotiation is subject to the ever present condition of elsewhereism in the sub's work. Some jobs have to be done almost immediately (roofing and rocking). The starting of other jobs has to be closely paced with the construction (plumbing and cabinetry). The duration of the job also must be settled upon, especially when it determines when the next successive job can be started.

To judge the reliability of the subcontractor regarding his holding to promised time is difficult. Such a judgment revolves around elsewhereism's potential for interfering with and delaying the sub—because the patsy does not or will not compete well with the others who are making demands on the sub. The patsy must judge well, because several of his other jobs can be thrown off if this one sub delays his promised time. (If an electrician does not finish his rough in, the walls cannot be sheet rocked and tiled.) So how can the patsy trust him in advance? The referral system may instill trust in potential temporal compliance by a sub. If a job is simple and immediate, and the sub says he is not busy, then the patsy can feel fairly certain that the sub will show up on time: If the job must be paced into the construction and is complex, lasting several days, then a higher probability exists that elsewhereism will delay the job. The problem for the patsy then becomes how with effective strategy to ensure the promised time? But how can the patsy be sure that the promised time is not merely the sub's ruse to get the job by agreeing to any timing; and then once it is tied up in a contract, just do it whenever it can be fit in with work elsewhere?

The patsy has several ways of ensuring promised time before choosing a bid by using his edge in power symmetry. First of all, he calls the sub as far in advance as possible, to minimize the effects of elsewhereism by planning ahead of time. The danger in this strategy is that the patsy may err—when the sub says he can do the job fast—by countering that there is no rush, there is more time. In essence, he lets the sub off the hook, and is liable to be put aside in favor of demands of a more pressing nature from general contractors. Then, of course, work elsewhere may amount to actual delays for the patsy. Thus the patsy must time his advance call to be within the realm of need so that he does not lose, either through a slip or the construction not being ready, the eagerness of the sub to keep his promise. The sub can easily risk the patsy's contentment in order to keep a general contractor's account.

At the site visit the patsy tries to schedule the subcontractor firmly, if at least tentatively. In pressing for a firm promise, the patsy insists that a job has to be done on a certain date or number of days. In return for this firm promise, the patsy may talk "as if" the sub's compliance with his timing means he will be chosen for the job, and/or the patsy may promise immediate payment as reciprocation. Both these latter strategies by the patsy are good "bait" for getting a firm promise. Obviously they can backfire by leading the sub to promise anything in order to get the job, but then letting nature take its course. The tentative schedule is based on a "span" promise: the sub is given a span of time rather than a specific time in which to do the job, once he is called. "As soon as I call, you can do the job within ten days." This minimizes the potential interferences of elsewhereism by leaving the sub a span of time to articulate his various clients' demands and to turn to the patsy's job.

The patsy also may probe the sub about his elsewhereism in order to make a judgment about how he (the patsy) competes temporally. He asks such questions as, "How many jobs do you have going? How many jobs do you have before me? Which jobs do you have to take first?" This is one of the few times he has access to any detailed knowledge of a sub's elsewhereism. Probing can also help the patsy determine whether the promised duration of a job is realistic. He asks the sub how long it took to do similar jobs. As the sub is leaving the patsy typically confirms the promised time by repeating it to the sub who again verifies it as a promise. Finally, the patsy may put into the contract a time clause to protect himself against delays. The clause may provide for breaking the contract or penalizing the sub in some way in case of delays. For single home construction by patsies, however, such clauses are rare since many contracts are verbal and it is easy to break even a written contract before work has started.

Subcontractors have various strategies for handling the demands of the patsy with respect to their other demands. They become either completely agreeable or hedge their work with an elsewhereism reference. The agreeable stance indicates complete understanding of the patsy's articulation of the total job and their desire to facilitate his construction.

This is a common stance since it plays on the patsy's desire to have his job understood and facilitated. The sub underlines his agreeableness with a ready eagerness which sounds like he will drop all other commitments and come right over. "Call when you are ready and I will come immediately," he will say about the job that must be paced. When the job is ready to go immediately, he says agreeably, "We can start this afternoon or tomorrow." When it is necessary to buy materials, he will say, "All we need is a day's notice." When the patsy asks about the duration of the job, the agreeable answer is, "It will only take about . . . we like to get in and get out fast." It is up to the patsy to decide and probe for whether the job is too fast for quality or reasonableness. Also, the sub's agreeable stance always leaves open the question as to whether the sub will keep his promise—after securing the job with it. The patsy soon learns that elsewhereism's ever present demands will interfere with many (perhaps half) of the promised times. However legitimate the interference is—and often it is not, as patsies do not compete well with generals—the patsy learns to hedge his promises by uptiming his needs in order to reduce the effects of potential delays.

Elsewhereism references are made by the sub both as explanations and apologies for not being completely agreeable to the patsy's pacing. The typical elsewhereism reference is that the sub asks for advance notice before giving a firm promise; he then adds, "Boy, am I busy!" He intends to give the impression that the patsy will get a fair shake in the competition with people elsewhere. Also, he may indicate his flexibility if he can have advance notice by telling of the number of crews at his disposal, implying that one is almost always free. Indication of priority is also frequent: "We squeeze the new construction jobs in right away," said a roofer whose repair jobs are as frequent as new roofs.

The apologia for delays in promised time is typically, "I have so much work now that it is hard. It is either feast or famine." The patsy then realizes that in a period of feast it is hard to nail down a sub. Some subs flatly refuse a definite time with the "bunched up" excuse: "If I could have got the stuff done today that I should have done last week, I could have got to you tomorrow." These subs are not trustworthy on timing, since the patsy clearly has no temporal control over an impossible schedule.

A negative elsewhereism reference—suspect when building is strong throughout the county—is the disclaimer of other jobs: "I hope we get this job because we really need it, as work is slow now." The patsy can easily assume that, although denied, elsewhereism lurks as a very possible delaying factor; otherwise the sub is not to be trusted on quality because he is not getting jobs when they are plentiful. The completely busy sub will simply admit it, if he has not already during the initial call, and either ask the patsy to wait until work elsewhere permits or place him back in the referral system with another competent sub.

After the laborious detailing, the arrangement of a promised time for starting the job comes quickly. The patsy must respond fast or lose his temporal control to the sub who has commitments elsewhere. The following exchange with a plumber is an extreme example of how the patsy can be detailed in his quest for the proper time.

Patsy: Well, if we give you the job, how long will it take you to get started?
Sub: About three days.
Patsy: How long will it take you to finish?
Sub: Three to four days.
Patsy: O.K., sounds good.
Sub: Boy, oh boy, have we been busy. We've been hitting them. We're doing two and a quarter million dollars of business right now. We're doing two tracts out in the country.
Patsy: Oh, boy. I can't compete with those people!
Sub: Well, you know it would only mean a couple of days' delay, unless somebody throws another tract at us. But we have plenty of plumbers, we never get tied up that much.

Days later the patsy realized that much of what he heard was aggrandized and he could have tried to negotiate specified time into the contract. But because of the intense elsewhereism, which he could not hope to compete with, he did not trust the sub and looked elsewhere.

Some contractors give the bid and price when these are simple, at the close of the site visit; otherwise when they return more complex bids, that time becomes crucial to choosing the subs. The agreeable, sweet-talking contractor will emphasize that he will phone the bid in later or mail it that night, to be sure he is in the competition for the job. Some contractors bothered by elsewhereism will return the bid quickly, asking the patsy for a quick response, whether yes or not. Then, the chosen sub will arrange his schedule and promise a time. When the patsy wants a sub, working under these conditions, he complies quickly, hoping that he competes with clients elsewhere in order to articulate his construction at best a pace as possible. Other contractors plagued by elsewhereism will delay getting out the bid, and lose the patsy's trust in their ability to stick to a schedule. Of course, if the delay is due to other reasons, then the sub is just not reliable.

After everything is negotiated and subjected to appropriate strategies, the patsy still must accept that elsewhereism's emergent character will generate delays for many of the chosen subs. Usually, however, these delays although annoying, will not be too adverse to the pace of his construction. Delays here refer to a few days, not to several weeks. Those few days of delay will require continually to rearticulate, in some moderate measure, pacing of the total job; this is par for the course.

Now let us turn to the detailing of price and the build up of trust in the cost factor.

Detailing the Price

Once the specifications and temporal aspects of the job are detailed, it remains for the subcontractor to give a price for the job. Unlike the detailing which preceded firm specifications and timing for the job, the detailing of the price—how it was figured—occurs *only* if the patsy questions the sub, for whatever reason. The typical reasons are to see exactly what was charged for, to check for mistakes, and to verify sub-prices. In his turn, the sub takes a *justifying* approach in order to win trust in his bid, as he explains in detail the costs of the job.

The basic rule that subs go by is to give a firm price and never vary it except for additions, extras, mistakes or actual changes. One reason for a sub's non-variance is that he never knows how his bid compares to other bids for the job. This rule occasions pre-price strategies by both the patsy and the sub. The patsy tries to indicate what will be acceptable; the sub tries to make the price reasonably stick. And both are trying to forestall the necessity to change it.

Let us consider these pre-firm price strategies, and then turn to discuss the detailing of the price, and then what happens when a firm price cannot be given.

Pre-Price Strategies: The principal strategy of the patsy is as *I* have mentioned, to get the closest possible bid and to avoid "extras" by using site, not plans, bidding, and inviting the sub for a *site visit* at a stage of construction that maximizes his seeing what needs to be done and under the conditions which he will have to work- If still there is doubt about the closeness of the bid to reality, the patsy can ask, when appropriate, that the sub sketch what he is planning—the fireplace, cabinets, type of doors—to be sure the right job is being priced out. He can also ask to see samples of materials to verify if the work will be correct, such as the roofing insulation or the type of tile or hardwood planking. At this point he is also verifying specifications. Proper sketching and materials add to his trust in the subcontractor. These requests of his come before a firm bid, but can also be used after it to detail the price.

The patsy's next strategy is to give *polite warnings.* The most obvious warning is to mention clearly that the sub is bidding against others: "We are getting some other bids too." Thus, the sub is on notice that he is in competition with others, so that his price cannot take advantage either of an ad hoc monopoly or a vacuum—as the perennial story of subcontractor gouging would indicate does happen. Even if the patsy is "tired" of bidding out jobs or is sure he wants a particular sub and wishes just to one-bid a job, he warns the sub, (On one house this occurred with the hardwood floor contractor, who did a special job: he thought he was bidding all his fellow contractors, while we were sure it had only to be him—at the right price.) The patsy uses this tactic more after he learns ways of checking out one sub without comparative bidding, as through the referral system, pricing materials, asking other patsies what price they paid, or a phone call bidding for price ranges. Or, he knows prices do not vary significantly for such a job.

Another warning that the patsy may give is his price range, or a statement jokingly referring to not going too high, since it allows some respecifying of the job before a bid is given if the price, indeed, would be too high. This warning can also be followed through with an indication that the price will be a check against standard materials and labor costs.

A third strategy is to give the sub some kind of *bait* for going easy on a price that could be higher. One only has to say a high price can result in the sub's not getting the job, since he seems to understand it the best—this is often true enough not to be a ruse. Another is to remind him of the referrer, the referral system, or that it was a callback—that is, there is more business for decent prices. Another is to offer a piggy-back job—or another small job later (let the roofer hot mop in skylights)—or another big job if it works out either on the same house or another (e.g., more tile or masonry work.) (To a mason we said, "We'll let you do the whole job if the price on this one is good.") Still another good bait is to offer prompt payment from the building loan, since it is worth money to get paid quickly for many a sub. Indeed, many subs give a 2% prompt payment discount.

Lastly, the patsy may try to put the sub explicitly on the hook concerning trust, as opposed to letting trust develop implicitly. Thus the patsy can say: "O.k. Gordon, I trust you because you trust me and I know that you will give me the best possible price." The sub is then put on notice that his price as it comparatively measures will be the basis of generating or reducing trust. The patsy can further add for strength, that he will feed back into the referral system the trust he discovered through price comparisons. In this system, it is typical knowledge which subs are "too high" and which give "good prices."

In giving his price, the contractor, if he says anything at all, usually gives a *justifying rationale* which may take many forms. If he says nothing, he gives a flat, firm bid with some form of gestural pre-bid strategy. To justify what is too high a bid, perhaps even outrageously high, the sub has several leverage points. He can point to the style of the house, call it first-class, and say that it's worth a quality job that costs a little more. The addendum is to warn the patsy about cheapening his house with the standard bid. He can offer a preamble about the high-quality, perfect job that the patsy will get and, though the price may sound high, actually that the patsy is getting a break. These subtle comparatives can boggle the patsy until and unless he compares out bids. The sub can snow the patsy about his forthcoming reputation as a builder who demands the best, even if it costs a little more (just in case the patsy wishes to sell his house or build another). Another ruse is to say that the job is worth the price for the residential area, if indeed not also the house itself, particularly when the job is visible. The subs may profess to have kept the high price as low as possible by close bidding, even when—"we figure bids very close"—they obviously were sloppy in detailing specifications and therefore were cushioning the price. The sub may also explain the high cost by rises in materials and labor or changes in code requirements requiring more to the job, while acting as though he were helpless against this trend.

In contrast, preliminary justification of a coming low price is only *apparent,* as it may end up upon comparison just as high as the acknowledged bidder. One tactic is to preface an elsewhereism explanation for the low bid to secure a job, either because the sub needs just one more job to make the month profitable or because he is getting no jobs and has to have something to pay his crew. (All subs who used this bidding tactic never were, in fact, lower than the flat bidders and one was the highest.) Another ruse is to acknowledge the bid as an economy bid made to save the patsy money. These bids often are very low, jeopardize trust, and are an insult to the quality standard of the house; they also engender many extras later on to raise the quality when the patsy discovers the true nature of the job. The bids basically are an effort to land the job first, then increase prices later. Also there is the sincere low bid made to save the patsy money by cutting down unnecessary aspects of the job—the surveyor only did half the required contours—or letting the patsy participate (e.g., work at labor, keep books, or order his own material to cut costs.)

Using low or high bid strategies does tamper with the trust built up to that point, because most pre-price strategies sound like preparations for disclosure. Therefore, many subs will just gesture their calculations, if they do them on the job, and wait to justify the price after giving the bid if it is needed. The major gestures are to measure out the job carefully in silence and to calculate with a business-like face in silence. Another gesture is to "eyeball" the job and just give a firm price for a job which is standard and simple (such as tiling a bathtub shower, cutting trees, or roofing.) These subs give the appearance of knowing their exact costs and for what they can do the job unvaryingly; this evokes confidence. The patsy can only discover how much in line their prices are later, through comparisons, since obviously the subs *can* vary the price for the standard job after sizing up the patsy.

If the eyeball price is given for a complex job, the suspicion is immediately that a cushion has been built in—there is extra profit if the job does not go away. For example, "The bid will go up $30.00 if the truck cannot go up this hill." If for any reason the job may vary through difficulty or changes, the flat bidder may add "with time and material for variations" to cover his firm price and allay suspicions about cushioning.

Detailing the Price: Once the price is given, the patsy either questions it or the sub leaves. The patsy may, of course, phone the sub later to question details while he is comparing bids. Questioning prices forces a *justifying* explanation upon the sub's part, because of the rule about not changing a price once given. The sub must explain properly to maintain and/or instill trust without backing off. He must reflect his competence by showing how his calculations sensibly resulted in his figure for the job. Changes are only permissible by a change in the detailed conditions of the job which in turn necessitates a recalculation. Thus, the patsy may change materials, leave out a portion of the job (a raised hearth for a fireplace), or put in labor himself to reduce a bid. He may also ask for an added "touch" in the job which increases the bid through an extra (add a tile combing to a tile bathroom floor). Whether the patsy questions politely or directly, the situation and subs will vary only in the intensity of response. The sub still is being asked to make his price sensible, which he tries to do by detailing it.

A patsy often will not ask for detailing. There are several reasons: he may feel shy. He may not want the sub anyway. The price may just sound right on its own, or it may be comparatively within the range of other prices received for the job. But even if it compares right, the patsy may still wish to know exactly what he is getting for his money—for comparisons. The patsy's tendency will be to detail prices of complex bids, rather than simple ones—hardwood floors. It is in the complex ones that costs get buried and cushioned, as opposed to simple ones where it is obvious where the money goes.

Subcontractors use various ways to justify as they detail their prices. Their fashion is probably based on a personal compatibility in dealing with the particular patsy (because we found no sub, no matter how he justified, who would budge from his firm price without its stemming from necessitating change in the job itself). Maximizing the conditions for possible backing down did not erode prices. The aforesaid conditions were: pointing out obvious "cushions" and "fat profits" and insistent pushing by two patsies for a reasonable price. The sub handled the patsies with no wavering.

The standard mode of detailing price is *carefully to itemize* the labor and material costs for each detail of the job. This method is quite effective since, unless an error or change is found, the clarity of the expert sub can easily render the patsy mute. When clarity tilts the balance of power in the sub's favor, the patsy's principal remaining offset move is to opt for comparative bidding. The clarity of the itemization does help the patsy make comparisons when all bids are in. The sub, who may lose leverage when his itemizations are too clear for the competition, then tries to muffle them or rough them out while talking of the quality behind his work—a non-itemizable item. Clarity of itemization also helps the patsy decide where a job can be cut, if necessary or feasible, to bring a price into line. Subs who plan a reduced job as a fail safe measure use such clarity to cut costs. Clear itemization allows analysis of high or out-of-line costs so that they are understandable to the patsy.

There are two general sources of errors. One is a mismeasurement or miscalculation of large or complex jobs; the other is figuring a bid from the detailings of an estimator. Both sources can be either quite genuine and/or can be a strategy mix. The outcome can be more trust or distrust in the sub. The former occurs, if, say, the sub genuinely admits his error by showing the patsy his working sheet calculations and how the job went wrong, and by correcting the price accordingly. An erosion of trust occurs if, say, the sub makes no effort to correct or adequately change the price for his error. The latter pattern is linked with the desire of some subs to make mistakes favorable to themselves and not wishing to take the job without such a cushion. There is a delicate balance of trust at this point, of correcting errors, with consequences for power symmetry during the potentially ensuing job, which leads to blaming another "in depth" if the sub figures from an estimators measurements.

When itemization is too difficult, annoying, or clear itemization risks a price, another standard strategy that subs may use is "cost blaming." The cost factors of a job are materials, labor, working conditions and profit; the sub picks out one factor and blames it for the apparently high price and a justification. This apparent bulge of one factor may be accurate, but it does not gainsay that the sub may be trying to conceal the protective cushion or excess through another factor. (For example, one sub said, "Well, the union just went up on April 1 to $14.00 a day so we have to pass it on to the builder," which was true enough, but upon further probing for items we discovered a profit in excess of the typical 10% profit and 10% overhead. But being caught at such concealment did not break adherence to the norm of never changing a firm bid. To justify this cushion of profit, the sub referred the patsy to the competitive bidding system as if to say, "You'll see, I'm not too high.")

Cost blaming is a way of concealing cushions pertaining to the sub's unexpected costs. It deflects attention from the cushion, while clear itemization directs attention to the cushion. The cushion is not merely a result of greed. It may clearly be a result of facing unknowns in the job that might raise costs, so that the contractor must protect himself; and/or it may be the result of "losing my shirt" in a previous job so that the contractor fears being hurt again. If the patsy gets far enough to discover a cushion, his trust in the sub will vary as to how well founded the cushion is as a protective device against a real probability of excess costs. If well-founded, he patsy can either accept the cushion or take his chances—after some negotiation—on doing the risky part of the job on time and material or allowing the sub to charge him extra. Again, trust in the sub—its development to this point—will be the principal desideratum on which alternative the patsy chooses. To try forcing a reduction in the cushion with no escape valve is unlikely to work, in view of the norm of firm prices. But if it did, or if the patsy discovers a sub who clearly had no protection against increased costs in his itemization, the probability is that the sub would either do a low-quality job or run out on the job rather than lose money. "I'd be working for the owner, not myself," the subs say.

The sub's other strategies of justification weave into itemization or cost blaming. One strategy is to follow up with suggestions toward either getting comparative bids or doing the job himself, if the patsy thinks it can be done cheaper. Another is to suggest the patsy do some labor for the job—if possible—to reduce the sub's costs and thereby the price. Discounts may be offered, as on materials or on prompt payments (2%), but which compared to the total price usually mean little change in price. The sub may suggest price reduction through changes in the job or materials, which by implication justify the original price. He also may suggest taking a chance on time and material or a lower bid with possible extras. These suggestions and miniscule offers give the impression of working with the patsy to help meet his budget which, hopefully for the sub, engenders the development of trust.

No Price: Several conditions may make a firm price unadvisable, as in the work pertaining to many expert-layman relationships. The principal condition is that the job cannot be planned—it can only evolve. Another condition is having to price out a job despite no visit to the site. Thus, to put a price on the job is either impossible or would require an excessive cushion against changes, extensions or problems. This occurs in masonry and tile jobs when the patsy must see, as the first condition as the job goes, both what he has and where to take it to—such as designing a fireplace. It occurs on such jobs, also, as putting in underground work where the terrain is questionable. In these cases, the patsy is presented with a time-and-material price, and perhaps a "guesstimate" of about how long the job will take and about how much material will be used. These guestimates give limits to probabilities of price ranges. Even this is a lot of information to go on, compared with taking one's chance with doctors and lawyers on problems of an unforeseen and unplanned nature. The second condition occurs when the patsy must have a quick price over the phone to keep the total job moving—there is no time for a site visit.

A sensible, well justified "thee and me" price engenders trust in the patsy, since the effort is to keep costs commensurate with the job. For comparisons, the patsy can seek other "thee and me" prices with guestimates. If he finds a sub who will give a firm price under this condition of an unplanned job, he is likely to mistrust the sub as giving too fat a price or not being able to do an adequate job for the money—which means trouble.

In short, the subcontractor arrives at the site to detail specifications, timing and price. In so doing, his effort is to build up the patsy's trust in him and in his ability to put out a quality job within the promised time for a comparatively reasonable, and certain amount of money. Insofar as the sub succeeds through some combination of his own efforts and comparative default, he is chosen for the job. In the next chapter we turn to the processes involved in the patsy's comparatively choosing the subcontractor.

IV

Choosing a Subcontractor

The comparative bidding of subcontractors, we have seen, is a three stage process. Chapters II and III have taken up the first two stages: inviting bids and detailing jobs. The third stage—choosing a subcontractor—is composed of two joint processes: *comparing* bids, and *handling* subcontractors while comparing. Comparing bids can go on while inviting and detailing bids or may await the time when all bids are in. In either case, comparing bids is a decision-making process which takes *time*. The pacing of bidding during this time may range from being quick to prolonged at points for any one job. The principal conditions of variation in pacing are the temporal articulation of the total job as well as the demands from elsewhere on the subs involved. The variation in pacing requires that the patsy hold off pushy contractors, delay others, while quickening the bidding of yet others as he compares and makes his choice. Thus, as choosing becomes more imminent, the handling of people becomes more pronounced.

Handling people at this stage can easily become more difficult a task than comparing, which mostly is a fairly straight forward calculation. In keeping control over the subcontractors while choosing among them, the patsy clearly has the balance of power symmetry in his favor as based on his power of choice. His only potential loss of this edge is when a subcontractor's bids elsewhere are accepted, as these occupy the sub's work force. The patsy may have to act faster so as not to lose the sub whom he wishes to choose.

It may be noted parenthetically that most analytic renditions of decision-making processes are based on the calculative activities, not on the people handling that is involved while decisions are being made. To be sure, some description of how people are controlled, especially during the moments of intensification when choice is drawing near, accompanies these analyses; but this descriptive level is seldom raised to an analytic level wherein the structural processes of how people are handled does yield theoretical variables that affect the decision outcome—just as the calculative variables do. The main point I wish to make is that the style and success of people handling feed back as determinants of choice; and the patsy, in our case, clearly builds this into his calculation. For example, pushy contractors may erode his trust, while the friendly compatible contractor, emiting patience and understanding easily may offset a cheaper bid by gaining trust.

In the pages below, I begin with a brief discussion of the comparing of bids and its calculative aspects. (Much of this calculation has been discussed in the previous two chapters.) The remainder of the chapter is devoted to people handling processes: more particularly to how the patsy handles subs while choosing, how he handles subs how are trying to "beat" the system, and how he terminates relationships with subs who are *not* chosen.

Comparing Bids

Comparing bids is a two stage calculative process. First, the patsy compares the bid on the detailings of price, time and specifications with respect to the variables (discussed in the previous chapter) regarding each detailing. In making these comparisons he compares price initially, then he compares the quality of the bid with the best price to the quality of other bids; he may also compare the relative understandings of the job specifications by the subs. If the best price is linked with sufficiently high quality and understanding, then this is still the most favored bid. If not, a slightly higher priced bid, with a better quality more suited to the patsy, may be more favored. Then this favored bid is compared to the other bids with respect to the starting and duration of the job. Again, if the most favored bid is out of articulation with the total job, then another bid may be chosen. However, the patsy may attempt, using his power of choice, to get the timing he needs. (I shall discuss below how he uses this leverage on the sub to try to obtain the best starting time.)

The second stage of the calculative process is comparing the subcontractors on the degree of trust that the patsy has in them. This stage may occur during comparing of details or when the subs have been ranked on the most favored bid. Trust is compared on specification, price and timing, regarding the degree of confidence that the sub has developed in the patsy pertaining to adherence to his promises. To be sure, the sub proffering the most favored bid may not have won the trust needed for its acceptance. But usually trust will cluster sufficiently with a favored bid for it to be chosen.

Comparing prices indicates quickly whether they converge or spread. The ideal range is a slight spread between a high price and a low one. This brackets in the appropriate price, and gives the patsy confidence in choosing the right bid when the qualities roughly follow suit. If the prices converge too closely with varying qualities, then the patsy may invite another bid to get some spread in order to bracket in the "right" price. He may simply get a rough bid on the phone as a check, or may go through the whole process of detailing on site. Contractors, too, are concerned about whether their bids are within an appropriate range of prices and a few ask to see if they are in line, "Does it look about what you wanted to spend?"

"Gouging": Sometimes the price spread is quite wide for the same or near quality jobs. Then it is clear with the high bids that the subs are "*gouging.*" They have built in excessive cushions or do not particularly care about the job unless it really pays, or are bent on getting excessive profits from the patsy. When the sub is trying to gouge with high prices, the patsy is protected from it by comparative bidding. But all high bids are not necessarily gouging. If the bid is hundreds of dollars higher it is gouging; but if it is $15.00 to, say, $50.00 it may only apparently be gouging. What it actually is, is calculated pay for a place in the system of elsewhereism—in which general contractors are making claims on the sub's time and giving him work over a period of years. If the sub now is going to commit himself to a patsy, and not be drawn elsewhere for a good account, then the patsy has to pay something extra for his work. If building is slow for generals, then subs relax their prices. This is the patsy's fate in this system of competing claims. His best recourse is to try to build during the winter when building activity is slower.

There is also low price gouging, through the "economy bid" for a job of less but seemingly acceptable quality. This job leaves the patsy open to excessive, costly extras, perhaps enough to throw out his whole budget, requiring cuts in other jobs. Thus, as the economy job proceeds the patsy discovers how "cheap" it is, and may pay to upgrade it. For example, as the meaning of an economy cabinet job becomes clear, he may reason that it will wreck the quality of his entire kitchen. To discover earlier if an economy bid is really adequate, the patsy may have to get more bids. He also must query the sub on prices of potential extras: e.g., more shelves in cabinets or better plywood. At first, the economy bid is hard bait to resist. The patsy soon will learn, however, through comparisons and a few jobs finished that the best economy comes with a realistic price for the quality offered. Otherwise, "trouble" becomes costly in time, quality and money. Some subs, realizing the potential danger of their being defined as offerers of economy bids in an effort to win trust, will both detail the prices of low bids and any extras that may occur.

Overruling Factors: The criteria for choosing can switch to other factors if it is a toss-up between two favored subs, or if a new factor overrules the standard calculations. These factors are basic qualities of people and go beyond, yet feedback support to the buildup of trust. The degree of personal *compatibility* with the sub becomes an issue at either extreme. Having a sub on the job day after day requires a measure of men being able to get along together. An obnoxious sub, no matter how good and trustworthy, can hamper the work of others on the whole project. (Two subs were nauseatingly solicitous and slick, another was rude to the point of distraction. They would have been disastrous to deal with and have around. Few actually were incompatible, more often subs fit the other extreme—highly compatible. They appear more than normally enjoyable and helpful to have on the job. For example, the electrician was so helpful that he got the author's bid, and indeed made the project at his stage a joy because of his pleasant company and good ideas about building.)

Feeding into compatibility are the factors of previous *friendship* through work or sociability, and previous evidencing of *skill* and *experience* that are highly desirable by another job at another house. Sometimes the patsy has seen the other job and sometimes the sub tells of his previous work in an entirely convincing manner.

One of my jobs, the cabinets, was so *complex* that it was too exhausting to go over the details more than twice for the bid and only one of the two bidding subs had mastered the complexity sufficiently to be trusted. In order not to lose the protection of the comparative bidding process in the face of his kind of overruling factor, the patsy can "false bid" the favored sub by telling him before the price is given that the patsy will call him in a few days after more bids are taken. This stimulates the rendering of competitive bids.

Elsewhereism can overrule the bidding process, if either it draws subs away from site visits or those who do submit bids must know quickly yes or no because of demands elsewhere. Then the patsy is forced to grab a bid, before he loses all the subs for the time being and so puts his whole project out of temporal articulation.

Other factors that may overrule standard calculations are the trying of somebody new or the spreading of work around when subs converge on price and quality. (The roofers converged so I tried a new contractor, relative to previous jobs who was also an old friend.) Another factor is who will do the work, the sub or a crew? If not the sub, is the crew experienced enough? (One excavator was cancelled when I discovered that an inexperienced crew rather than the very experienced sub*, was to drive the front loader.) Yet another factor is the possible leverage for favors gotten from the sub for another job: for example, the door contractor provides a load of pipe for a water line at his own cost, saving the patsy considerable money.

These overruling factors—and there are more—are not used by the subcontractors to beat or get around the bidding system. They emerge as part of the patsy's calculations. When we consider beating the system tactics, these are actively introduced by the sub as strategies to overrule comparative bidding, and become more a people handling problem feeding into calculations.

Non-comparable Bids: Calculations become difficult when bids are non-comparable. This condition allows the other factors discussed, as well as system-beating tactics, to swing the bid and perhaps erode the protection afforded by comparative bidding. There are four related sources of non-comparability of bids: the complexity of the job; the degree of the whole job the sub will do—partial or complete; redesigning information gaining and additions in successive bids; and lumped vs. highly detailed bids.

Unlike simple bids, it is difficult to keep complex bids comparable. For each successive bid there is much room for suggestions, redesigning and additions and subtractions as based on information gained. Non-comparability increases at a fantastic rate. Even apparent comparability may easily break down to non-comparability when the patsy later discovers that each sub detailed things that the other sub passed over, given the room for discretion and mistakes in complex bids. It becomes very tiring for the patsy to keep pressing to keep bids comparable, and of course it is impossible when additions to the successive bids are made. Bids also become non-comparable in their lack of, or difference in, listing extras. The patsy then is faced with interpolating the bids, and calling subs later to find out how much difference in price there is for an addition or subtraction. In the end the patsy can only hope to achieve a rough comparability, yielding a general idea of what he is going to get for each bid.

Sometimes the patsy is stuck with virtually complete non-comparability when a complex bid comes with the details lumped as a whole job—“rough-in electricity” or “per conversation”—versus another bid which details exactly what each item is that will be roughed-in. (Past experience swung the author’s bidding choice in favor of the detailed specification bid, though it was higher in price. The lumped bid engendered poor trust, and phone calls could not overcome the severe lack of details.)

Phone calls are only good for picking up prices on older bids and for additions on new bids, especially for bringing into line simple bids—for example, calling back a dry wall contractor to check price changes for rough texture walls. Phone calls also help for checking on price and reviewing details for understanding. Subs are usually quite willing to clarify, as it means that their bid is in the running. This also provides them with the opportunity to make small adjustments with price and to make their bids sound sensible. If brave, the sub may also sound out the patsy on his problem and help make the bids comparable, such as explaining the difference between four and three-and-a-half ply roofing.

Non-comparability resulting from partial and whole job bids requires that the patsy make up the difference with how much it would cost him to complete the job or to hire someone else. Usually his saving is more apparent than real. In bringing the bids into comparability he tends to fudge on his own costs, and then discovers that it would have been cheaper and better to get the sub to do the whole job. On virtually every job there is some part the patsy could do himself. He can grout tile, lay water line, hang fixtures or rain drains, but . . . experience if not leading to skill, which it often does, does lead to speed which maintains the pace of the whole job. Doing it oneself can be overdone on a temporal basis, and thus may raise costs, since time costs money even for the patsy.

The subcontractors themselves often worry about comparability of bids—they do not want to lose a job for the “wrong” reasons. So some will give their partial job bid as it is, detailed after redesigning, and then give a bid on the whole job. Or, if asked for a special job, they also bid as they know the standard sub does. Others will ask the patsy to be sure his bids are comparable, to guard against the sub who gives a non-comparable bid in order to make it look better—a “beating the system” tactic. (For example, the tile contractor pleaded with us to make sure others gave the more costly, mortar job bids: “Some guys go low because they glue on the tiles, you know, and this does not last.”)

Handling Subcontractors

As the calculation proceeds, the pacing of the bidding process requires, as we have said, that the patsy hold off the subs moving too fast and quicken the subs moving too slow. With the power symmetry edged in his favor through ability to choose, the patsy has leverage for controlling the subs while bidding, comparing and choosing.

The patsy paces subs within the temporal limits set by: (1) the articulations of the total job. which may force waiting or faster bidding. (2) the intrusions of elsewhereism and (3) the temporal limits of bidding contracts, which give usually 30 to 60 day limits on choosing the bid before changes can be made.

Pacing—Holding Off: Holding or putting off subs is a fact of life in bidding. For every sub that says "Call me when you've decided," another will say something pushy, "Do I get the job?", that indicates he wants to know immediately if he can "go ahead." And yet another, through his quick efficient bidding, as impressive as it is, will generate a condition of pressure on the patsy to respond in kind which would be too early in the choosing process. This juncture-of receiving a bid and not answering—requires delicate politeness and certainly no abuse. The patsy has the power to put off choice, although the sub may become the chosen one, and bad feelings easily arise in these fateful situations.

Built into the response to receiving a bid which is not taken immediately, is the normal strategy of indicating that one is holding off the decision until bidding has come to a close: "Thanks John. I'll call you when we decide which bid." If the patsy is actually not bidding a job, this statement also serves well as "false" bidding. (Subs who think they are not being bidded—perhaps because they are dealing with a patsy, otherwise I am not sure why-responded to such a statement with a disappointing grimace and silence. Having given a fair price for a decent job, they had expected to get an immediate "go ahead.")

In addition to the normal strategy, three others emerge quite easily in response to the current tenor of the relationship. If the patsy is friendly with the sub and really wishes to choose him, but wants to complete bidding first to be sure, then he *encouragingly* holds him off. "I know yours will probably be the better because I always give you a $50.00 edge, anyway," or "I am almost inclined to say go ahead and do it but I better wait. . . . I mean, there is a 90% chance I'll call you." Should the sub not subsequently be chosen, then this encouragement can result in a sour future relationship.

When the sub gets pushy, the *consultant* strategy easily arises, as if to say the patsy chooses him but first must confirm it with others involved. "I must check with my wife, it is her home," to which the sub is likely to say, "Sure, check with the war department." Or, "I must discuss it with my brother first." Then, of course, such expert consultants can be referred to as architects and building inspectors. This put-off still allows the sub to think that he is in the running.

Lastly, the *timely* strategy readily occurs to the patsy when obviously there is more work to the bidding so that it will be some time before choosing can actually occur. In response to a lump price on a complex bid, "Well, you better send me the papers on it. I haven't decided yet, but in case I choose yours I want to just be able to sign the contract and get you started." Or, "Well, I'll think about it, it's actually two weeks before the job has to be done anyway and all bids are not in." This strategy is used particularly when the sub attempts to put the patsy on the spot to decide fast—perhaps trying to make him jump the timing of the normal bidding. As a counter strategy, the patsy must communicate, "all in good time."

Pacing-Quickening Bids: The speed with which the patsy gets a bid is an important indicator both of the sub's eagerness for a job and of the elsewhereism that will have to be contended with in any sub. Some subs delay bids to hold up bidding, as ways of gaining time to see if they would rather go elsewhere or just because they do not wish to be chosen. Others are just slow. But most bids are delayed for definitely structural reasons. A principal one is that a complex bid takes time to work out and write up; another is that subs are "busy"—demands elsewhere take their time. If these conditions are combined (as they are for many complex bids such as with plumbing, heating, electricity and doors) the patsy can easily find his bidding delayed. He waits for the bid in order to make his choice.

At this juncture if he wishes to quicken the receipt of a bid he makes phone calls "to see if the bid is figured." The response is invariably based on elsewhereism as a legitimate excuse: "I've been snowed under today, and I just didn't get it run out. I'm working on several bids, here, right now and yours is next." At these moments the patsy realizes how hard it is to compete with general contractors for the sub's time. Sometimes the sub will add up the bid while the patsy waits on the phone, or he may call back—unless he forgets for elsewhereism—but he never will let his big accounts down for a patsy. Though the patsy may get a price over the phone so that he can make his choice for a complex bid, the chances are slim that the written contract will follow immediately. It comes only when the big accounts have been taken care of.

At other times the sub will say that the bid will be "in the mail today." If the patsy then asks, "Well, then what is your price?" he is as likely as not to get as an answer that the sub has forgotten and can't the patsy wait until tomorrow's mail? This clearly indicates that the bid has not yet been done and that perhaps it actually will get in the mail if meanwhile a more lucrative account does not call the sub from elsewhere. One sub, guilty about the delay, said he hadn't totalled up the bid yet, "because I am trying to find ways to cut corners" for the patsy; this, of course, was a "cover." (He was a door contractor who specializes in doing apartment houses, so he found a home a comparatively very small job, hardly worth the effort. His bid actually came out low—a quality economy bid.)

Also, the patsy is likely to run into a protection barrier when phoning the lagging sub. A secretary may answer "who," when she finds out the patsy's name, or will say that the sub is out (whether he is or not) but will call back. She will know nothing about the bid except "He was working on it, but had to rush out"; again, elsewhereism is the excuse. If the patsy gets upset he may be passed on to a second, strong protection agent, such as an estimator, who will tell the same story plus that he knows nothing about the particular bid. In the face of these barriers, the patsy can only wait or use what little power he may have—power diluted by elsewhereism—and leave a message for the sub that there is the deadline for bids or say something to the effect that if the bid is not in the mail within a period of time, such as today, he will assume the sub does not care to submit a bid. Ultimately, if these deadline pronouncements do not work, the patsy either waits or goes on with the choosing of a bid. Some subs will call back within a few hours and delay the patsy even more by giving a low but partial bid or guestimate, to be further calculated in a "few days." This bait is likely to entice the patsy to go on waiting and hoping the actual bid will be as good. He soon will learn, though, that it is not worth holding up the total building project on such an expectation. The time and trouble of putting off schedule other jobs is not worth the savings of the low price bid—if it does come through, which it probably will not.

Pacing-Confirming Leverage: When the patsy has made his choice, he must disclose it. Within the act of disclosing he can use his confirming leverage over the sub—he will confirm the bid as taken for a slight advantage or change. Usually price and quality are fairly fixed, and this leverage is used for maneuvering a favorable change in promised time—either in starting or in duration of the job—or emphasizing the importance of adhering to the promised time. ("If I take your bid, will you start next Tuesday and be out by the end of the week?" was the leverage used on the chosen electrician. I asked him if he could start immediately, and he said "sure." Or, an adherence, "If I give the bid, will you start telling me your trouble about other jobs, and drop me to work for someone else?" This particular sub responded, "No, I can handle your job on schedule," thus reaffirming his promised time.) Whether this reaffirmation is true or fudged is a judgment that only the patsy can make. Then he must live with it. If the patsy still has his doubts, he can promise to call back his confirmation later in the day to give himself time.

Sometimes this confirming of promised time ferrets out a delay in it. In the meantime the sub has taken other jobs, and, if honest, responds that he can no longer do the job when planned for: "I won't be able to do the fireplace this week because I'm really busy." Then the patsy must recalculate and negotiate a promised time, while trying to ward off yet further elsewhereisrn in the next promise. However, for this sub the patsy's trust in adherence to promised time will never be firm again.

Often the patsy must act fast in order to offset the "first come, first serve" credo of adhering to the promised time of bids. When he hears, "I can't let other people down, you know, because they are all waiting," he realizes that for busy subs, he must get in line and hope this morality of elsewhereism applies to him, too. (In the next part of this book—Generalling the Job—I shall discuss what happens when promised time is not adhered to.) The patsy never really knows whether after choosing a sub, promised time will be cast to the winds or "rained out," so that his building project will be delayed. He can only try to use this last bit of leverage before taking his chances, and then handling whatever delays occur. If he pushes too hard on confirming time, the sub may start immediately but then disappear for a time: he has hooked the patsy who now is forced to wait.

The confirming leverage also can be used to request a contract that guarantees a job, and which makes sure that the sub will get the job passed by the building inspector, as with plumbing and electricity. But many more jobs require guarantees, such as paving, roofings, floors, tiles, and formica. Since most subs will just go ahead with a verbal O.K., the patsy may hold off confirming until a contract is produced and signed. This is more likely to force a recalcitrant sub to stand behind his work for the written contract is better and more efficient protection than the general law that licensed contractors must stand behind their work. A contract also provides the sub with protection concerning being paid on time, and thereby facilitates use of the mechanisms of the lien on unpaid bills.

Terminating Bidding Relationships: Once a subcontractor is chosen, the question arises as to whether other bidders must be told if they were not chosen. Most subs do not require a response unless they are chosen, although they often are interested in knowing who got the bid. Some will ask to be told "Call me either way", and others will call back in a few days to see who was chosen. Thus, frequently the patsy must say "no" to subs, and then the problem becomes whether or not to explain, and if so how? Those who wish to be called may be using the call as a way to push getting the job. Also, they like to know how they are competing and may be juggling their promised time commitments. A "no" allows them to commit themselves elsewhere. On his side, the patsy soon learns that the bidding system rule is not to call back unless asked to be told "either way." The less talk the better, if only to save his time and energy.

The pacing of a "no" is also important. The norm is that the announcement should come after all bids are in and a choice has been made. This norm can have untoward consequences for the late bidding sub. If before his bid is in, the patsy chooses another sub, then the late sub is wasting his time working up a bid. Yet he cannot legitimately be told before his bid is in—as it would not be fair for him not to have been in the running—unless his lateness is excessive or clearly he did not meet a necessary deadline. As a result, from time to time subs sent bids too late to be in the running and have to be told "no." Thus, the late sub has worked for nothing. In this norm the bidding system works well. But its pacing generates this potential waste for slow subs or for subs occupied elsewhere although still trying to make a bid. (In one case, I broke the rule, and told a sub that his bid was too late for the choosing. The sub became furious and said, "I got the bid out in five hours." It was a complex foundation bid. The patsy had to remind the sub that, however, it took him three weeks to get to the work-up. This response indicated that breaches in this rule of saying nothing were not wise. It is best to say nothing and pace the "no's" appropriately.)

The easiest way to say "no" without pressure to explain is to phone when the sub is likely to be out on a job and leave a message with his secretary, answering service or wife. They seldom will ask for a reason, but faithfully take the message. Even the curious sub will find it hard to call back for an explanation, whereas he might easily press for one if he had been on the phone.

Explanations must be gracious to avoid rancor and not prohibit future work relationships. Perhaps the current choice will not work out and to finish a job the patsy will need a sub to whom he had said "no." A gentle, vague fib works well, when the truth would be offensive. Instead of telling a sub his price was "outrageous," he is told that he "was in the running but we wish to try . . . something about timing. . . ." (and mumble). The patsy also can say that his wife wanted another subcontractor so what could he do! Sometimes the sub will provide the explanation when asking for it: "What happened, did the wife say no?" The patsy need only agree in order to leave the relationship congenial.

One sub, in an effort to nail down how he is competing, regularly sends out a questionnaire inquiring into the details by which the choice was calculated: "want job," "want job, but at a reduced cost," "changed mind and do not want job," "the sub was not low and gave the job to someone else," were four beginning statements to be checked. The questionnaire went to the details of detailing. A self-addressed envelope with stamp was provided for a return. The patsy still responded graciously so as not to offend.

Beating the System

The problem for the subcontractor who does not wish to submit to comparative bidding, or who wishes to do so as little as possible, is how to prevail upon the patsy to give up the protection which comparative bidding affords. How can he engineer a *break down* in the system or invoke criteria of choice which *overrule* normal bidding calculations? If the former, he can push the normal conditions of bidding to the breaking point. If the latter, he engages in interaction maneuvers. I shall consider both kinds of strategies, but beforehand I wish to make clear that this is *not* an overmanaged strategic view of the subcontractor. By this I mean that the sub is not consciously saying to himself, "How can I manage this patsy?" Rather, subs develop a style of businesslike behavior which is congenial to themselves and is constantly adjusted and corrected in the crucible of real world consequences. In this sense, not only is the purpose of his style to compete in the bidding system but also to gain a special advantage over his competition—even to overrule it. His style adjusts itself, as he succeeds or fails. Also, the strategies themselves are mild.

Breaking Down the System: We have already had examples of ways to push normal bidding activities to the point of breaking down the protective aspects of the system. There is the economy bid which admits of many extras later on, but might land the contract at a low price the patsy finds unrefusable. There is the monopoly associated with a geographical area. There is the detailing of a complex bid to the exhaustion of the patsy, so that he just takes the bid, feeling the cost of going through another bidding is not worthwhile. This is the patsy's false bid to keep the sub's price down: "We are getting other bids. There are the delaying tactics—not showing up at a site visit or submitting a bid in an effort to halt the bidding process so that maybe other sub's will be drawn elsewhere, or the sub can get a job done elsewhere while still bidding the patsy's job. A low guestimate or promised visit or bid, works in this regard. Also, there is bidding for extensions of a job—as if it were a waste of time to get more bids since they already are on the job (the tile contractor's suggesting kitchen tile while doing the bathroom, or the mason suggesting path work while doing a fireplace). Perhaps they even get the patsy to go on time and material, and give up a price as protection, using his trust in the current job. And, of course, bidding just becomes a bore at times and the patsy may not want to bother if he can check out a bid another way. Having now discussed how the patsy handles these pressures on normal conditions, we turn next to the strategies which subs use to overrule normal bidding.

Overruling the System: There are many ways to try overruling the system. They are primarily focused on playing upon the patsy, the sub himself and the system. Most modes pertain to the patsy—the friendly approach, hooking and clienthood. The *friendly* approach is grounded in sweet talk and is resolved by development of obligation in the patsy. His indebtedness is paid by the contract. Sweet talk, as we have seen, is the way of subcontractors; they try to be nice, jovial, understanding, supportive and helpful. They sincerely wish to render service. They try to solve problems and provide helpful hints. They offer a modicum of friendship. They are very interested in the building project, and flatter the patsy on any number of its "lovely" aspects. They never correct any of the patsy's statements for fear of offending him, unless to do so is absolutely necessary for a detail on the bid. They do favors, perhaps, like buying something at cost with their resale license—providing the product has nothing to do with their own job, and sometimes even then. The friendly approach can backfire resulting in distrust, if the bid clearly is a "gouge." Otherwise, this approach does win trust and can easily work if the patsy "buys" the sincerity. If the approach does not quite overrule the bidding system—as it did for this patsy a few times—it can easily win in a toss-up comparison of bids. Some subs can act so successfully at this approach that they deliberately and overtly assume they will have the job. "Here is the man that is going to pave your road," the sub introduces himself.

Various ways of *hooking* can overrule the comparative calculations. An effort is made to promise something special that the patsy wants within or in addition to the bid itself. Within the bid, the promise can focus on a special detail, a price or material rebate or a desirable promised time. The more complex the bid, the more potential for hooking on some special detail. The effort is to use one detail to beat out all the other comparable bids. For example, one electrician offered 200 feet of extra cable for getting the bid to save the patsy the expense of a power-pole. Whether the sub actually carries out his promise is a matter of trust and judgment while choosing and verifying and perhaps of negotiation while the job is in progress. If the hooking works, then the patsy may not even bother to finish bidding and will calculate a choice, feeling that the added specialty tips the scales completely. One roofer offered to "hot mop in" skylights a month later for nothing, but the patsy did not bite. However, these inducements are hard to resist, when trust exists. Yet, as we have seen (and will further discuss in the next part of this book) the sub can easily forget a promised time or a promised desired quality.

Another strategy that subs use is to try to socialize the patsy into being a client. *"Clienthood"* means that the sub becomes the expert (much as if he were a physician) who is simply expected to take over the case and perform competently. The client then does not comparatively bid his experts; his loyalty and trust lead him to yield this protection and control. He becomes vulnerable to some exploitation but does not see it as such—and probably is not exploited. He is merely turning a job over to an expert who does not have to compete, and who will take care of it without bothering the patsy. The client becomes a passive good "account." These efforts failed with the author as patsy and with other patsies whom I have interviewed; socialization into clienthood works well, and they remain "patsies." The sub becomes the expert in charge, and any doubt or discussion on the matter is disloyalty. Some subs do win solid clients over a period of years by proving themselves; but this is quite different than trying to convince the patsy to be a client right off.

Several less frequently used strategies involve playing on the sub himself or other subs: the direct sell, elsewhereism, the sob story, squealing and beating out others. The *direct sell* is when the sub implicitly discounts the bidding system and his competitors, and just sells himself in some way. He may give a long favorable rendition of his history. "I have 170 accounts and only lost two customers in the last seven years." This particular sub then switched tactics to socializing the patsy into clienthood. (And he succeeded when the patsy heard some of the names on his list of accounts. They were the best of general contractors.) One sub gave a persistent, long, but soft sell on price. He went over his job and all alternative jobs with a fine tooth comb. Then he sold the utility of the job. Some subs try to sell themselves as a monopoly among all subs; only they have a certain material or piece of equipment (a type of furnace or carpet), or special knowledge of the patsy's special problem. Subs who remodel older homes will bill themselves as specialists in taking apart certain kinds of construction with special care before remodeling. We are familiar, of course, with the *elsewhereism* strategy. The sub says he is in between jobs and can do it now if the patsy catches him quick. The implication is to take him immediately and forget comparative bidding.

The *sob story* is hard to resist because someone is being saved. The sub tells the patsy how he needs the job just to keep his crew, his business or himself, going. He may even say that his price is "bare bones" for business. He may combine this approach with the friendly approach. In a toss-up among bids, he is particularly hard to resist. (But thanks to comparative bidding, all the sob stories that I met were proven to be devices for getting the job. One particularly heart-rending story narrated from a liquor breathing mouth—"we're nearly broke"—proved to be a cover for an outrageously high price for a mediocre job.)

Both *squealing on* and *beating out* other subs amount to breaking the rules of the bidding system. Squealing is warning the patsy about another sub contractor who cheats on materials or does very poor quality work, and who is likely to be bidding the same job. Squealing as a transgression is justified with, "I do not mind bidding against reputable subs, but I am not going to lose a bid to a guy like that." Thus the squealer is a "knight" come to the rescue. He deviates to save the system, not to wreck it. In contrast, the sub who "beats out" other subs is clearly hurting the bidding system. His approach is "Tell me your lowest bid and I'll beat it." It is very tempting to enter into collusion with this sub, since there can be no recourse except perhaps negative referrals if the patsy later should build another house. This approach, of course, breaks both the secrecy norm of the system and the norm against collusion in fixing a price. (Fortunately for this author, the request came only twice, and the morality problem was easily resolved in favor of the system's integrity. The offending subs were judged as untrustworthy.) Thus the bidding system is supported by a very high majority of honest subcontractors. On his side, the patsy dares not use this unfair strategy, for most subs would be quite offended, leave immediately and put out the word to all subs to steer clear of him. This could severely hinder his whole job. The patsy then can be controlled while an individual sub may get away with this strategy for beating the system, since it is hard for the patsy to make the sub's action visible to other subs.

Lastly, some subs *take off* on the system itself. They may simply pooh-pooh the system: "Penny-pinching bid getters drive me out of my mind." This is an effective strategy with people who have sufficient money for any size of bid, and feel guilty about "shopping." They do not realize that comparing experts *is* the legitimate way to get the job; as it is not, in many other realms of expertise. Other subs will take off on the system by obfuscating their bid with bewildering interpretations. "I can see you like to work yourself. This bid will be low and you can provide some of the material and work with my man so you do not have to pay for an extra man!" It sounds good, but is hard to figure out. Only comparative bids could really put into relief what the sub had to offer. (Another one said, "I am going high because I am including everything, but I bet this is still your lowest bid for an expensive house like this!" It was momentarily hard to resist the comparative interpretation for a quick go ahead, but upon later comparisons it proved just an average bid for a standard fireplace.)

These tactics for beating the system are by-and-large ineffective. The one which can tilt close calculations sometimes is the friendly approach. In sum, the comparative bidding system, while apparently easy prey to all kinds of breakdowns, injustices and transgressions—because of its uncodified, informal norms—is surprisingly immune to such occurrences. My only explanation is that sticking with the system provides the best protection—job for price—for both the patsy's current project and for most subs over the long run. This system works, therefore, before either side tries to evade it; they use it fully, get what they want and end with no reason to evade it.

Now that the subcontractors have been chosen through comparative bidding, we turn next to the patsy's *generalling* of the jobs themselves.

Part II

Generalling the Job

Generalling the total job, as noted in Chapter I, refers to: (1) comparatively bidding the subcontractors; (2) administering the temporal orderings of the job, which entails supervising each subcontractor over time and articulating many jobs simultaneously and sequentially, and (3) inspecting, evaluating, trouble-shooting, and hence controlling the quality of work, and then paying for it. The discussions in Part II will be about temporal and quality control and about cost-payment of ongoing work; as the power symmetry shifts back and forth in the patsy- subcontractor relationship, particularly when the patsy's edge is being undermined by elsewhereism. In the typical expert-layman relationship, the expert will general the work to be done. Here we have the opposite: the layman will general experts who in turn either do their respective jobs themselves or run their crews who do the jobs.

The patsy's prime goal in generalling is to gain close control over the cost, timing and quality of the various jobs, hence to maintain control over the total construction. This is the best way for a patsy to secure the house that he wants at a price he can afford. Another goal of generalling is the fun and satisfaction it provides the patsy. This is how to make a future home exactly what he wants as he grows it *and* grows with it. It becomes a central joy of his life—building his own home.

Of course, there is much to do in generalling besides the managing to be discussed in this part of the book. There is finding the building lot and developing the site, designing the house and having plans drawn, securing financing and a building permit, working out a division of labor and choosing a season in which to build. These facets of generalling have been discussed briefly in Chapter I.

Finally, to engage in the controlling of time, quality and costs, the patsy must be able to act with confidence when facing the subs. If he becomes scared, he thereby may easily abdicate his control because then the power symmetry shifts in favor of the sub who then proceeds to run the job according to his own temporal convenience. Thus the patsy must have the ability to assume the several attitudes and stances of the general contractor in order to keep the balance of power symmetry sufficiently equal. Generalling, then, does not come easy for the typical layman; for he is prone to accept his know-nothing status, hence to relinquish near-complete control to the experts.

So, in Part II, we shall see how the patsy fairs in generalling. In Chapter V the temporal control of a job is discussed, and in Chapter VI the temporal control of several jobs. In Chapter VII, the quality control of a job is discussed, and in Chapter VIII the closure of a job.

V

Temporal Control of a Job

The central issue in generalling the temporal aspects of a job is to keep it on schedule. This means to get it started and finished on time, so that both the next job can begin and the course of construction itself can be kept on schedule. Thus, the central problem in generalling temporal aspects is coping with malpacings of jobs. Most occur because of *delays.* Only infrequently do they occur because of speed-ups by the subcontractors.

Delays in starting and finishing are normal. A subcontractor who arrives and finishes on time, to do a job when promised, does occur but he is not to be counted on to do so. The reason for the prevalence of delays is that their potential is a ready consequence of a combination of three principal conditions of the total construction project under which the patsy must constantly general. Of these three, the most prevalent condition is *elsewhereism.* The subs or their crews are constantly being required elsewhere, and the patsy usually competes poorly in his demand power (power symmetry). Secondly, the *course of construction* seldom proceeds for a patsy as planned by him or promised by a sub. Both unfulfilled or inaccurate expectations for timing and actual hold-ups of a job generate the criteria for defining when a delay has occurred. Thirdly, as elsewhereism takes its toll and the course of construction is slowed or changes, the *articulation of successive* jobs is thrown off by a chain reaction which yields yet more delays.

In this chapter, I shall discuss briefly why delays can be more or less consequential. Then I turn to a fuller discussion of the sources of delays, followed by consideration of how both patsies and subs handle delays. (The articulation of several successive jobs will be discussed in the next chapter.)

Consequences of Delay

Delays in a job fall on a temporal continuum and have varied consequences. Sometimes short delays may have drastic consequences when they hold back other jobs—as when the roof is delayed a day, then it rains the next day and this in turn prevents the beginning of jobs which require a dry, covered house. Usually the longer the delay, however, the more drastic the consequences. Short delays are bothersome but not disruptive of work, since there are often a few days of slack time after which, only then, if a job is not started or finished, does other work become disrupted. Indeed, the patsy will tend to warn subs before he actually readies for them, so that short delays will be absorbed by lead time. Thus, the electrician is told well in advance that he must have the rough-in work done by a certain date, so that the walls can be closed in with sheet rock.

It is, of course, their consequences for disrupting work that make the managements of delays so crucial to the patsy. Even when a delay is having no virtual effect on the total project, it can be demoralizing for the patsy to face an unfulfilled promised timing. The delay puts into question the sub's reliability, hence the patsy's trust in the sub. It sets up a warning that the job can really go awry, unless one closely watches the sub and his timing of the job. The patsy must general such a sub every minute, or accept the consequential disruptions to the project. In contrast, when a job starts and finishes on time—especially if on promised time—the patsy feels satisfied because the total job is advanced and the next job proceeds on schedule.

Delays have multiple effects in four areas: articulation of successive subcontractors, over-all timing of the total job, quality of work, and the patsy's private life. The most important impact of disruption is in ruining the articulation of the timing of successive subcontractors who will appear on the job. Delays can set off a chain of sequential delays in many successive jobs. In turn, this necessitates re-articulating the jobs of many subs. This is time-consuming and often is difficult, as promised times for starting and finishing must be rearranged and reintegrated among subs. These drastic consequences easily occur when the roofer or rocker is delayed by rain. Also, the difficulty is increased when the patsy runs headlong into commitments and competition from elsewhereism, because he has lost his priority through postponement of a job. He must re-enter the competition against demands from elsewhere at a disadvantage because of a sub who cannot be depended upon to be ready for a job at a time when it was promised. He loses considerable edge in the power symmetry needed to rearrange a favorable timing with a sub.

The patsy has a perception of the time that the total job should take, and on this basis he plans for how long his building project is allowed to take. He plans also, on the basis of this cutoff date, to move into a new home, give up another home, go to another job and make closure on his financing arrangements. (If he plans to sell the new house, he also plans the percentage profit on this timing.) Thus, although it is infrequent, any prolonged fundamental delay can be most disruptive—say, the house is finished a month or two too late. To forestall such undue, prolonged delays which would have this consequence, the patsy must outwit the rains, and if necessary, cancel out a procrastinating sub and get a new one, for whom a delay does allow time for. He also must be sure that all jobs will meet inspection requirements, so closure on the total job is not prolonged by trying to get a sub back to re-do an inspection failure. In one case a mason did not, according to the inspector, put in enough wall tabs on a brick wall. It took three weeks to get him back to put them in—his crews all being tied up elsewhere at the other end of the county. In contrast to the inspector, he felt the job was o.k. and he had earned his money! This delay would have prevented a close-in of all walls and brought the job to a hall if the inspector had not, under much demanding, allowed the remainder of the house to be closed in and proceed ahead of the finishing of this wall.

Starting delays can easily occasion rush jobs in an attempt to finish on time anyway, either so that the sub can meet a commitment elsewhere on time or because of pressure by the patsy. Rush jobs, in turn, can constrain putting the necessary time into a job that will allow quality—especially if the sub thinks the patsy will be none the wiser about the lowering of quality. Duration delays—the job taking longer than planned for because either of its nature or other external contingencies—lead to rush jobs or to not finishing properly. Again quality is reduced. (For example, the heating crew ran out on the job without flashing the chimney, since they took too long on their other work. Then it rained, and seeping water stained the redwood ceilings. Another instance: an excavator spent half the allotted time fixing his tractor, so did a terrible rush job on the excavation.) Delays which end in low quality work, force the patsy into sometimes heated, but always some, negotiation to bring the job up to standards without an increase in price, even though the sub may have run out of contract money and have to do it on his own money, thus run a loss. (Jobs that have been malpaced by the infrequent speed-up, rather than a delay, may also amount to rush jobs with the accompanying poor quality.)

As for the effect of delays on the patsy's private life: it is two-edged. Short delays with negligible consequences for articulation and the total job can prove to be welcomed respites from the tedium and virtually continual pressure of the ongoing job. It is nice to have a day or two off. Unfortunately, more often delays can leave the patsy frustrated and unsatisfied, and if prolonged, quite demoralized. He also may become quite bothered by all the rearticulalion required from prolonged delays which often take much time from his personal life, and require cancellation of social engagements or other appointments. The job must go on, and it is up to the patsy to general it: if he does not, nothing happens. As one sub sympathetically put it, "Every day you stop cranking, it all comes to a halt!" Even when the job goes well, satisfying as the work may be, it takes its toll of wear and tear on the patsy; but delays compound the toll, without reward.

While delays usually are caused by subs for their own mainly favorable reasons, some delays can cause subs to lose money and jobs. As other jobs pile up, the subs may easily receive cancellations and/or a barrage of angry customers. Meanwhile they may be losing money on their contracts if a delay involves the duration of a job—and they may create a dissatisfied customer if they give him a rush job of low quality.

Sources of Delays and Elsewhereism

There are many sources of delays, and the patsy can never be sure when one is going to be operative. Some sources are, nevertheless, *expectable* (that is, he can expect them with some certainty although not knowing when) and others are *unexpectable* (that is, he has no grounds for ever expecting them). The expectable sources of delays are many and most involve or belatedly relate to a facet of elsewhereism. Thus, although the particular source may come as a surprise—a crew being dragged off in the middle of the patsy's job, delaying it, for a quick job elsewhere—it is to be expected, given the patsy's inability to compete with general contractors. The expectable sources, while temporally unpredictable, are frequent and readily understandable. Their patterns allow the patsy and subcontractor to have a repertoire of strategies to cope immediately with them when they arise. In contrast, the unexpectable sources while unpredictable too, are infrequent, and understandable only after some probing—why did the material not arrive? Unlike the expectable sources they are not patterned, and the patsy may have to develop new strategies on the spot to handle them.

Expectable Delays: Elsewhereism intrudes in several direct ways to delay the patsy's job. The demands of general contractors elsewhere are often more than a match for the patsy in competition for the sub's work. The patsy can easily lose the balance of power symmetry with the sub when such demands are made upon the latter. The patsy either starts off with low priority or loses through circumstances what ever priority he has over the sub's time.

The patsy loses to elsewhereism whatever priority claim he has on the sub's promised time in several ways. If the date is arranged but the patsy does not have the construction ready and asks for a day or two delay in starting, he can easily lose his turn and then have to wait out the demands from elsewhere before getting a new date. The same can happen when the patsy fails to warn the sub the requisite day or two in advance that he is ready for a job. Again, a few days of local rain can force a pile-up of roofing or excavation jobs with general contractors around the county, and the patsy cannot hope to compete with them—so his job is left undone for days, which throws the total project off for many weeks. Even with the roof on, his house must dry out before finish jobs can be done, and weeks of rain are depressing. This is the particular danger of building in the winter, although that is the best season to avoid the competition of the general contractors.

The intrusive power of general contractors is clearly felt when a patsy is told the eve before starting that the sub unfortunately is delayed by a heavy schedule; or must take care of a contractor first (NOW!); or has been running late for weeks; or a job is taking longer than planned because of extras. Sometimes a sub is so busy he cannot round up a crew because they are spread over the county at various jobs. And sometimes, smaller subs who create crews just for each job may find their trusted men working for other contractors who are too busy to let them go. This flexible approach leads to many delays in congregating a crew; it is used particularly by small excavation contractors. Often when the sub is heavily demanded he just goes, does not phone the patsy, and fails to arrive to start. The patsy then must track him down, only to hear what he expected—he has lost his priority to more important jobs.

Inept message takers can lose the patsy his priority when phoning in for the job. While bungling the message can delay the job somewhat, the subsequent consequence typically is that the patsy loses his priority during busy times to jobs elsewhere; so the latter eventually accounts for the extent of the delay. If not, the sub would resolve the bungled message within the day because he wants the work.

Of course, a major reason that patsies lose their priority over even a promised time, is because they so frequently have such a low priority to begin with. (In the next section I shall discuss how the patsy offsets this drawback with appropriate strategies.) One indicator of this low priority is, as noted previously, that many subs will promise any starting time to get the bid and then service the patsy when convenient, with respect to demands for other jobs. In short, these subs assume this low priority and the relative powerlessness behind it, or they would be more careful with their promises. In doing the job on their own temporal terms, they pace the construction. They do try to meet the patsy's needs while ignoring his demands, but only if the demands do not interfere with meeting the more important requirements of general contractors. The bigger contractors especially do this (for jobs such as plumbing, masonry, heating, electric, excavations and drywall). Other subs not too closely linked with the pace of construction may just come when they can (door, glass and cabinet contractors). To the patsy this latter practice may appear subterfuge and delay, but the canny sub sees it as the nature of things and as helping out a patsy during busy times.

Another frequent indicator of low priority is the pulling off of a crew during a job to do another "little" job when a general contractor screams for it. All of a sudden the patsy will see the crew disappear for a few hours—before his own job is finished: the sub or crew will say "They'll be back in a few hours or in the morning" as they "have to do a little job" or "are needed to help finish a job." The patsy realizes at this moment how shifting the priority balance is for those in his low position. Many subs do this in order to make money and to take care of others, because the patsy is virtually powerless in midstream of work on his job. One extreme example: an excavation contractor charged for moving his bulldozer onto the site for a week-long job; but he was under such pressure to put in septic tanks elsewhere that he moved the bulldozer out every day at one o'clock to do other jobs, and moved it back in at 6:00 AM to do the patsy's job! The patsy was only charged one moving-in fee. In another somewhat less extreme case, a general contractor, himself, came to my job and literally pulled a crew away with him on orders from their boss—the sub—to finish up his (the general's) plumbing.

Also, a sub may phone during the job and tell the patsy' that he will not be back on the job for a few days, "I have to do a big job in Terra Linda," and leave the patsy hanging. The patsy again realizes that his ability to compete with generals elsewhere is low.

If a crew is outlawing—that is, working for the patsy's sub while being laid off a few days by another—he may be drawn back by the original boss right in the middle of a job. This occurs somewhat through loyalty, but mainly through expectation of continuous work for the crew. Again the patsy realizes his low priority in that he too was just a squeezed-in job for the crew and sub while they waited for another sub or job to get working again. In all these examples, the patsy discovers that people elsewhere can have prior claims over the sub or crew, which they only have to exercise for a delay in his job to occur.

Unexpectable Delays: When expectable delays occur, the patsy and his commiseraters (usually other subs) often say, "That's the way it goes." But when the unexpectable delays occur, then the patsy has "bad luck" and if many occur, "a run of bad luck." As said earlier, the nature of these delays is "Who could have predicted it?" If these delays had been expectable, usually some planning could have forestalled them; whereas it is impossible to forestall elsewhereism from the patsy's position of weakness.

Perhaps the most understandable kind of unexpectable delay, "after some explanation," derives from equipment breakdowns. The equipment may be old or poor, and simply break down under the strain of its job, or the crew may wreck the equipment during the "danger hours," that is, after 3:00 p.m. when fatigue so often results in errors and mishaps. (This happened to me particularly with the excavation and roofing equipment.) Then the patsy must wait out the delay caused by the amount of time it may take to get replacement parts and to fix the equipment. The duration of delays may range from "being back on the job in the morning" to "keeping in touch" while dragging out interminably.

Lack of materials also can delay a job. Special materials may be hard to find; a material may be out of stock, or a bad estimate may leave the sub out of supplies before a job is finished. Other unexpectable sources are that the patsy may fail to have the construction adequately prepared for a job, thus have the job postponed by the sub—and lose his priority—while preparing it. (This occurred with heating vent installations, tile counters and masonry footings, which required detailed preparation beyond the author's knowledge.) The sub may give an unrealistic promised time—to get the job or because of poor judgment—thus making the patsy's temporal planning run into a string of delays and making the time spent on the specific job appear to be a delay.

During a job subs and/or their crews also may simply take off for vacation, short trips and personal problems. Crews may quit their subs for various reasons in the middle of a job, thus walking off the latter, and the sub must put together a new crew to get the job going again. Inspectors also cause unexpected delays. For example, if one unexpectedly does not make a close-in inspection as arranged, a whole series of jobs get delayed. Or, an inspector may saddle the project "out of the blue" with a stop-work order because he wants something done, undone or redone, feels this work must be "forced." (This occurred once during the foundation work done for me. Out of a clear blue sky, a policeman arrived with an order to stop all work until more reinforcing steel was installed—beyond what was accepted in the permit. The patsy was also threatened another time with a stop-work order after a completely spontaneous inspection of ceiling heights and masonry work.)

Finally, the sub can halt work a day or so by failing or forgetting to secure sub-permits, as for excavation, plumbing, septic tank connections or electrical work. For example, the excavator was going to work all weekend but forgot to get a license on Friday, so the next job was not ready to work on Monday. Accidents and power failures also can cause delays but seldom occur.

Handling Delays: The Subcontractor

As expectable delays occur, usually due to elsewhereism, subcontractors generally make no effort to handle them except by the most perfunctory of rescheduling. They assume the patsy knows what is going on. They may even avoid his weaker queries as to "what happened?" by not answering the phone, and having a secretary or crew member announce the reschedule. But when the patsy's queries are strong with anger, insistance, threats of changing subs or demands to fulfill a contract by a promised time, or when the delay is unexpectable, then subcontractors make a direct effort to handle the patsy.

They have an armamentarium consisting of two kinds of strategies: *reduction* and *hooking.* Reduction strategies are designed to quiet down the patsy, reduce his anger if necessary, make him happy, and keep him "off the sub's back" for as long as possible. At the same time, many of these strategies have the purpose of keeping the patsy "hooked" to the particular sub. If a reduction strategy does not have this consequence too, and it appears necessary to keep the patsy from going elsewhere, the sub will follow with a hooking strategy. Giving a new promised time is the epitomy of a condensed reduction and hooking of the patsy. The patsy is shut up and left to wait again by agreeing, however tacitly, to the new timing of the job. Now let us consider the strategies that subs use when the patsy is pushed or when pressed by unexpectable delays.

Reduction Strategies: The initial reduction strategy is typically some form of explanation which will bear upon the source of the delay and, if truthful, will usually point to demands made elsewhere of higher priority. Thus, once again, the patsy discovers he is either low or has lost priority in competition with general contractors for the sub's services. The delay is made sensible in terms of his ongoing-shifting but usually low position, hence his low power symmetry in relationship with the sub during the job. The truthfulness of explanations will justify the delay and thereby serve to maintain the trust that the patsy has in the sub, thus keeping the patsy hooked.

One form of explanation is a detailing of the *backlog* of commitments to others which have been occasioned by the rain or some other delaying factor. As the detailing goes on, the patsy develops the notion of how many jobs elsewhere have priority and when his turn might come. The sub may then follow through for hooking purposes—especially if the patsy is still annoyed—that he can *stall others* perhaps (much as the patsy is stalled) to get to the patsy sooner, thereby implying that the patsy's priority has gone up. The stalling, he might explain, consists of putting off another job that can wait more easily; or he will start another job, then squeeze in the patsy while the other job waits for him to finish. This "stall others" strategy makes the patsy feel better since now his priority value has gone up. As one sub explained, "I got a couple of other jobs to do right after this one I am on, but, I mean, I can stall them off a little bit to catch your job." Another sub said, to soften the delay, that "I am going to do enough to please the current job and then start on yours."

Another explanation is the sub's *delineating* of the articulation of his total workload so that the patsy understands his own new position in the lineup of jobs—which invariably will be favorable so as to keep the patsy happy and hooked. This *rearticulation* combines the backlog and stall strategies. (As the foundation contractor explained, "I have one more day on the house I am building. I could put off another waiting job, and start your work tomorrow afternoon." He then apologized for not starting "today.") Seeing the place of his job in the sub's juggling of many jobs quiets the patsy down, as that juggling seems quite reasonable.

Begging off the delay is another prevalent form of elsewhereism explanation. Basically, problems arise in the job previous to the patsy's and the sub gets stuck working them out; hence he delays the patsy. When the sub finally calls to explain, he apologizes, complains and goes over the history of becoming stuck to show that it was not his fault, but involved a typical circumstance which foiled normal planning. For example, "Well, the only thing I can say is I'm still not done down there. Man, I'm wallowing around in the mud—it's sure a mess down there. Ditches are caving in on me and everything else," explained an excavator detained by rain.

Blaming others is a frequent aspect of these explanations. General contractors wreck their schedules by goofing up their course of construction; crews walk out or make mistakes and suppliers fail to deliver on time; or message takers do not pass the word as soon as they should. Blaming others deflects the fault, so the patsy may be reduced to siding with the sub because both are the victims of delays generated elsewhere. In sharing the other victim's fate, the patsy is still quite hooked to the sub.

Of course, not all explanations have to do with elsewhereism in the world of subcontracting. As we have seen, there are other sources of delays that can be explained—such as why even anticipating rain can cause a delay in a roofing or masonry job, or why an illness lasted so long, or what the risks are in starting a job (such as plumbing) before the course of construction is absolutely ready for it.

It is not unusual for a subcontractor to come on with a massive assault, using all of these forms of explanation at once in trying to reduce an enraged patsy to patience. These assaults are hard to withstand and after one, the quieted patsy meekly said, "Thank you very much. I'm sorry that everything is so awful."

After the initial explanation, a second reduction strategy of a temporal nature will usually follow. The coup de grace after most explanations is following through with a *new promised time,* or with a promise to keep the patsy posted on when a new time can be given. "We'll get you tomorrow," or "I will call you tomorrow night and let you know how I'm doing here and when I might get away." And, of course, built into the backlog, stalling and rearticulation explanations is a temporal expectation as the patsy hears of his priority in the sub's series of jobs and the time that it will take to handle each. There are, however, other temporal strategies that subs may follow through with, or begin with if they do not offer an explanation.

The new time itself can be *temporally reinforced* to ensure reduction and hooking. Such statements as, "I am on my way right now," reduce the patsy to a happy, "Good, see you in an hour." Or the sub may say, "Don't worry, you are next on my list," after giving the new time. Again, "If I can get there sooner, I will," indicates to the patsy that the new time is an outside conservative figure so that if the sub's jobs elsewhere go well, the patsy may even get his work sooner than expected. This "immediate service" talk, of course, in cutting down on the potential delay, also eases the strain on the patsy's patience.

At the other extreme, giving a new time over and over, as each fails, can turn into another *delay strategy* to keep the patsy on the hook as long as possible, before he either gives up or the sub finally shows up. This form of delaying is particularly intolerable and erosive to the patsy's trust in the sub. It seldom occurs except under two conditions. Either the sub is under intense demand by general contractors and the patsy's job never warrants enough priority in the competition with big jobs and clients elsewhere. (This occurred with my plumbing, doors and cabinet subcontractors, who were doing subdivisions and apartment houses.) Or, the other condition occurs when the job must be repaired or redone under a guarantee to meet some failure or unfulfilled city inspection requirements. Thus, the patsy is hooked if he wants the work as it should be for no cost. And the sub needs only to agree to meet the guarantee demand but put off the patsy interminably while his crews are working on paying jobs. The patsy is likely to get the repair done after hours or before regular hours during a lag in the sub's schedule or during a rain storm—all times which are unpredictable. Thus, the sub just keeps promising new times until a crew can be freed at little or no cost to himself. Often this delaying is reinforced by the patsy's inability to reach an evasive sub in an effort to arrange a new time. As the delays continue, they are less and less justified and more and more met with silence or statements to the effect that there is nothing more to say except to give a new time. The patsy waits, power symmetry is clearly in strong favor of the sub.

Normalizing the temporal perspective of the patsy is another strategy that subs use to reduce the patsy's concern over delays. In so far as the patsy buys the adjusted time perspective, he continues to be hooked but quite happy. Normalizing occurs in various ways. The simplest is for the sub to assure the patsy that all jobs tend to get delayed; delays are normal irrespective of who is involved. Thus the patsy comes away feeling the competition elsewhere is not so strong and his priority is not so tow. Another way of normalizing is to explain to the patsy either that the job takes time ("Things don't just go that fast, you know") so the delay seems more a result of the patsy's perspective than of the sub's pacing. Or the latter explains that the job will not take as long as the patsy thinks, so not arriving on time will really not jeopardize the scheduled finish time, hence will not hold back other jobs or the total course of construction.

A more involved normalization of the patsy's perspective occurs with subs—plumbing, heating, electricity—who must gauge their work according to the *course of construction* so that they are not delayed. They are likely to promise the patsy any time he wishes, but then watch the course of construction carefully and not start until the building is completely ready for their job. Thus, they can go right through with it with a minimal number of visits. They are not likely to explain this to the patsy until they do not show up for the promised time, but simply query on the phone as to how the construction is going: "How is the house going?" When the patsy complains of being delayed, then they try to adjust his perspective to doing the job at the best moment as that will not slow the total construction down. When and if the patsy picks up this perspective of starting time varying with the right moment of construction, he quiets down. Validation of the adjusted temporal perspective comes with the sub's ability to synchronize well, otherwise his statement will appear as merely a delay strategy. But most subs synchronize very well; and as the inexperienced patsy begins to understand, his perspective normalizes to a "course of construction" orientation.

In admitting this to the first sub who normalizes his temporal error in starting time, he loses power symmetry over this job relationship. Then he offers to keep ensuing subs posted on the right moment to begin, rather than set a definite date which will only generate a delay. At this point he becomes better at generalling, which gains him balance in the power symmetry of relationship with the ensuing subs. He is now master of starting times again.

Temporal warnings are routine for many subs, who take a *fait accompli* approach to delays which undermine their trust with the patsy. They wish to be sure he remains reduced and hooked. These temporal warnings usually are occasioned by job trouble and delays elsewhere. Many vary from a vague notion of how delayed the sub will be ("It will be the middle of the week sometime. I'll let you know") to a clear idea, which varies from a few hours to a few days. (For example, the mason warned the author that he would be a few hours late for his job, and this warning was excessively careful of the relationship, because it had been somewhat less than completely trustful to that point.) Clear warnings are more ordinary. (The floor man pleaded, "I should be out of here by now; I started this job a long time ago, but I guess it will take the rest of today and tomorrow, so I'll see you the next day.")

Outside warnings are useful for giving the patsy the longest time that he will have to wait and an implied promise that the job will begin sooner: "We'll see you in two weeks definitely, and maybe earlier." *Near miss* warnings are handy in begging off coming immediately to do a patsy's job when the patsy calls: "If you had called me Monday, I would have done your job for you already." Implied is that the sub took another job and the patsy will now have to wait his turn.

Lastly, subs may engage in temporal action strategies either as a *set up* for impending delay or to *overcome* one. In the former instance, the crew works exceptionally fast and well, and then at the end of the day announces that they have to do another job tomorrow and will not be back until the day after. "But we'll be back to finish on Wednesday. We've just about finished now already, so we will not hold up other jobs while we are gone." To overcome a delay, the floor man hired an extra man to do sanding and moldings just to finish the author's job, which he could not get back to without more undue delay because of commitments elsewhere. Many subs use this strategy to finish jobs that their regular crews cannot get back to without the delaying of yet other jobs.

Behind explanations and temporal strategies is a composed manner of talk, carefully done, which puts over the sub's position while it reduces the patsy's annoyance. Virtually every sub has some contrived manner, which he has developed through time and because of trouble with many people. In dealing with patsies it is clearly a leveler of their excitement, and gives the sub the balance in power symmetry, since the more composed of the parties in a tense situation tends to have more control over it. The composed manner varies for different subs, the most common being "sweet talk." It is used mainly in giving explanations in a calm manner. When a new time follows, the sweet talk can turn into "happy talk," mirroring the patsy's joy and relief. Another way of putting over explanations is the use of double or imprecise talk which is impossible to fathom. There is also sympathetic and/or agreeable talk as the sub, in explaining his side of the delay, appears to take on the patsy's side too. Subs may also moan and groan about their problems to draw sympathy from the patsy, and thereby reduce and hook him. When pushed, or discovered in a slight misrepresentation, subs will go silent to squelch discussion and not to appear defensive.

It is hard for the patsy to resist being reduced by these composing manners of talk, since they deflate anger and make the patsy appear unjust or irrational to be angry in the face of them. Only as a history builds up in a relationship with a sub—when it becomes clear that the manner really is a subterfuge or is hiding the real source of delay—does the patsy openly get angry at the manner. Otherwise his counter strategies, too, take on a calm-rational lone, after he simmers down.

Also behind explanations and temporal strategies there can be an element of "Tell him anything to get him off my back" (as one sub put it), when the sub is not completely truthful. Typically the effort is to hide what is in fact going on elsewhere, so the patsy will not realize his low priority in the competition for the sub's time. Also, when subs are in a complete bind elsewhere and have no idea when they can extricate themselves, thus cannot tell the patsy something favorable, then they tend to "tell him anything." Usually this situation resolves itself into giving successive new times—and this becomes a delay strategy. This delaying is reinforced by the patsy who continually pushes for a time that the sub can meet. The few subs who chronically run late constantly use this strategy in an adept, agreeable manner no matter how false it comes to appear when nothing happens! They are adept at keeping the patsy happy and hooked beyond his best judgment.

Hooking Strategies: Besides the hooking aspect of most reduction strategies, sometimes subs may engage in direct hooking of the patsy so as not to lose a job, because of a delay and its apparent consequences. Another reason to hook the patsy securely is so that the sub can dangle him with many successive delays, thereby being able to take care of general contractors and jobs that will not wait. Securely hooked patsies can be dangled interminably while a sub is taking care of others elsewhere. (The landscape excavator, for one, regularly left the job at noon to do jobs elsewhere and thus doubled the time it took to do the author's job.)

The securest way to hook the patsy is, using *beginning strategies,* to start on the job and then to start pulling out on it in midstream to squeeze in other jobs. It is very difficult and costly for the patsy to change jobs in midstream. The landscape architect after settling on the plans took two months of successive delays to finish them. Another way, if appropriate, is that the sub may have materials delivered to the site, then let them sit and delay starting the job. Receiving the materials—for example several hundred pounds of sheet rock, or cement blocks or brick—incurs a responsibility on the patsy's part to stick with the sub, since he is liable to pay for the material through the contract. Another sub may not use the same material or supplier, or to negotiate the material away from the sub who ordered it is a nuisance and is time consuming. It is hard for the patsy to switch the power symmetry in the relationship by wresting the job away from a sub who uses these beginning strategies.

Other hooking strategies are less secure, but quite effective. Two temporal maneuvers are to *never give a time* for starting or continuing work so that it is hard for the patsy to establish the occurrence of a delay, except eventually on grounds that the whole job is suffering. This maneuver is used with jobs that need not be in sequence with others but are done by or near the end (paving or installing rain gutters). The second maneuver is to *keep the duration between successive delays short enough* so that the patsy can scarcely have enough time to get another sub. The danger to the patsy is that he might be bidding out the job when the original sub arrives to do the job. Therefore, the patsy must clearly unhook himself by calling off the job in order to rebid it.

Another is a *guilt strategy.* "We have a remodeling job coming up immediately but we can postpone it a week or so for your job." Besides being delighted now at his high priority, in spite of the current delay, the patsy would feel quite guilty at cancelling the job *now.* He is hooked.

Lastly, the sub threatens to walk out on the job in spite of his delays and this may reverse-hook the patsy. Instead of seeing this as a chance to get free, the patsy, at the thought of such a loss, is more likely to be quelled into patience once again.

Handling Delays: The Patsy

Despite all these reduction and hooking strategies that are used by subs to offset delays, the patsy is not without counter-resources, given the near balance in power symmetry of his relationships with subs. These counter-resources—such as the giving and the paying for jobs and the referral potential—allow him to use certain strategies to bring the delaying sub into line.

Expectable delays can always pop up, so the patsy must be on guard to handle the sub's delay strategies. No matter what were his initial strategies, the patsy soon learns to speak up about delays because of the stakes involved, and in doing so learns about other strategies and develops a readiness to use them. To accomplish this aim of getting the sub back on the job—minimizing any delay—the patsy's strategies help him to compete better with jobs elsewhere, to keep his power more level in the sub-patsy relationship, and to keep the sub actually coming (rather than killing the job altogether).

Most such strategies are pertinent to *elsewhereism problems* because this is the area where subs can change and bend the most, the area of flexibility and ability to rearrange jobs. By contrast, unexpectable delays, such as equipment breakdowns or lack of materials, are resolved fairly matter-of-factly in a space of time dictated by circumstances beyond the alterable control of sub or patsy. At best, the patsy can only try to hurry-up what normally, whatever the pressure, takes a certain time, such as repairs or reorders.

Basically, the patsy has three types of strategies for handling expectable delays: he can try to *ensure* the sub's arrival, he can *push* for getting the job done as promised, and he can pressure the sub into performance with intense *anger.* These types are ordered by the intensity of need to do something which builds up as the delay becomes more drawn out, progressive, or fateful for other jobs and the total job. Thus, the patsy may start with ensuring strategies, but if the delay is crucial he may get angry right away.

Temporal Ensuring: Ensuring the job temporally against delays starts during bidding, when the patsy negotiates for a promised time that fits into the succession of jobs. In resolving a delay, the patsy may immediately start to negotiate a favorable new time and hope to get the temporal jump on the sub, who doubtless will try to negotiate a time favorable to his other pressing jobs. Additional temporal ways to try to ensure a job are: first, to forestall more delays by giving the sub monitorings on the course of construction and warnings when to stock the job with material ahead of time, and second, to give a deadline, after which the patsy will put up with no more delays but will put the job out to bid again. (I gave only a few deadlines and never had to rebid a job.) Also, to request a weekend job is sometimes a good way out of a bind when competing with powerful generals elsewhere.

Perhaps the most useful way of ensuring that the job will proceed with no more delays is to play on the sub's sense of *responsibility,* which, after a delay is in question, is therefore more in need of reaffirmation to allay the patsy's doubts. The quickest way to make the sub responsible is to agree, accept, or otherwise indicate belief in his new promised time. An immediate response with no static to the new time, such as, "O.k. we will see you tomorrow, or at your convenience" puts the sub on notice that he is expected to adhere to his proposed, promised time. Any static, negotiation, or countering would undermine the solid responsibility indicated by the patsy's instant, complete acceptance. When appropriate, the patsy reinforces this responsibility expectation by repeating the date a few times—even several times. The same strategy works for tentative timing with regard to the course of construction. When a sub says, "Oh, yes, we can come up most anytime for a job like that," the patsy responds "Fine, I'll call you when ready," and says no more.

Adequate preparation for many jobs takes time, money and interferes with other work. To prepare for a job according to the sub's temporal request enforces a responsibility upon him to get in and get his work done as promised. Thus, putting a huge hole in the floor as requested for the fireplace makes it incumbent upon the mason to fill in the hole as fast as possible. Had the patsy put it in himself without a commitment from the mason, it might have been there for several days, or even weeks, as his own sole responsibility.

Lastly, to ensure that a delayed job will be done as rescheduled, the patsy may bait the sub with praise, money or referrals. Money, of course, makes good bait since no matter who the sub works for, he often has to wait a month or two to get paid. In the meantime, he often needs ready cash to meet his own expenses, especially payrolls. Therefore, quick payment of a completed job may readily draw the sub from elsewhere. (To add to the bait I once added, "I have the money now, but I have never been able to keep that much money more than a week, so you'd better get it while you can." To this the sub replied, "o.k. I will get it right away." He then pulled his crew off another job and tiled the shower the next day.)

Another form of bait is praise or flattery for the sub's crew. When the men take pride in their work, this draws them to an appreciated job irrespective of other mildly competing demands. And when the praise and appreciation of good work is followed with "I'm going to tell people about your work," there is usually no more delay. Crews especially like such praise, and pressure their sub to let them finish on time rather than being sent elsewhere before finishing.

Pushing: To further ensure that a delay will end, the patsy may engage in a pushing strategy which intensifies the pressure on the sub to get the job done. The patsy may apply straight pressure, use an insistent temporal strategy, or push by checking on the sub. The patsy must be careful in pushing, as we have said before, so as to keep the sub coming. Too much pushing can move him to the point of asking the patsy to get another sub or of sending a colleague to do the job.

Straight pressure is a direct attempt to level with the sub, while insisting that the job get done. *Need* provides good leverage for straight pressure. It may be an emergency call to finish a delayed closure to a job quickly (having the heating contractor flash the chimney immediately before a rain storm to prevent disasterous leaks). To delay this job would incur a liability on the sub's part: the patsy has to be careful to resist the sub's counter moves of sympathy—deflection and excuse—to get off the hook, so he does not have to pull a man from elsewhere for a special trip to the patsy's job. Other typical needs are to get a job done to obtain the next payment on the construction loan, so several subs can be paid or another sub allowed to start on time. These needs are quite vital and hard for a sub to ignore.

A *confrontation* also applies straight pressure on the sub. The patsy explains to the sub that he thinks he is being delayed in favor of others and being misrepresented regarding explanations. He asks the sub to level with him about what is really happening and whether he really intends to do the job. The patsy then may follow through with a threat to rebid the job. The typical retort by subs is to say that they will level, in a very sincere voice, *but* continue on with the same ruses. Subs never drop their posture or attitude, they only pretend to. Thus, this form of pushing gets the patsy nowhere. If from the start subs do not wish to disclose the real source elsewhere of the delay, they never will. For the patsy to push successfully, he must do so on the sub's own terms of what is real to him—such as arguing a vital need, bargaining for dropping a job elsewhere, playing on responsibility and dangling good bait.

Temporal pushing is part of and follows any negotiation for a new promised time to end a delay, and it goes in either of two directions: to clarify an obscure time or to press for an earlier time. Subs obscure the new time in order to generate delay by giving a promise without specific expectations and responsibility. ("We will come during the course of this week—middle of the week" or "Well, we are 19 days behind schedule now because of this rain." In both cases, I had to press for a clarification and for definite time ("I wonder if you know what day yet?") to offset the temporal uncertainty and ensure the responsibility.)

Once the patsy gets a definite time or continues to receive an uncertain time that can be bracketed, he may push for the job sooner, by insisting on and repeating how desirable it would be "if you could get it done earlier." Or he can quiz the sub on his schedule in order to offer suggestions and to reason out about how to squeeze in his own job. Or, he can apply the straight pressure strategies mentioned above. If he does not keep pushing for a better or a definite time, this will let the pressure off the sub. who then will allow the current obscurity or some later dale settle the issue, much to the dissatisfaction of the patsy. After a few of these experiences, the patsy learns to push unrelentingly or give up and accept; that is, he allows himself either to be or not be reduced by the sub's strategy.

To check up on the sub's activities is the most subtle way of pushing him out of a delay. An initial move is to constantly monitor the sub's activities elsewhere, to be sure he will get to the patsy's job on time rather than be waylaid elsewhere. To do this, the patsy makes phone calls to the sub's office and talks with anybody who answers, asking "What's doing?" Just talking is a way of pushing his own presence. Then, the patsy tries to forestall further delays by reminding the sub that his job is next or is to be done on a specific date. If the patsy foresees another delay because the sub is becoming stuck on a job elsewhere or because a new job has eroded the patsy's priority, he can try to forestall the coming delay through use of the ensuring and pushing strategies. If they do not work, and he is also rebuffed by the sub's secretaries, then the patsy may turn to his strongest strategy: intense anger.

Controlled Anger: The controlled use of anger is a very effective pushing strategy *after* a build up of delays and unfulfilled promises by a sub. To be angry too soon or before sufficient cause makes the patsy appear irrational, therefore he loses edge in the power symmetry. To become angry after sufficient cause appears justified, and few subs can tolerate it. Most will do the job without further delay, just to get the patsy "off his back." "The squeeky wheel gets the oil" subs say. Doing the job is virtually the only way to reduce the patsy and ward off threats of complaints to the contractor's exchange of negative referrals and of poor publicity, which most smaller subs who work on single family residences cannot afford.

The patsy's controlled use of anger must never get out of hand: if it does, he may appear as a "nut" however justified his anger, or he may completely drive the sub away even in the middle of a job. Therefore, this strategy is dangerous if the patsy is too angry to control himself. He need only act angrier than he is—not be that angry. The patsy has to be able to back off abruptly, reach an understanding and negotiate a new time, when the sub now wishes to do something. Too much anger may force a carry over that prevents this; again, then, he appears unreasonable. The patsy has to plan the time to "start screaming" strategically—not extemporaneously; he sets himself a deadline, paced with the build up of delays. He focuses his anger properly on the source of elsewhereism whose priority he wishes to undermine.

Uncontrolled anger lacks this focus. The controlled patsy sounds fed up and deserving, but not wild.

Basically there are two modes of getting angry: accusing or demanding. The patsy may accuse the sub of failing him ("You failed me Jack. If you sent your man up when you said you would, that chimney would have been water tight"), or of neglecting him for others ("By God, it hasn't rained for three days and you are not there working. What the hell's wrong with you? Who *are* you working for?"). The patsy may also accuse the sub of low quality work, inordinately high prices, non-cooperation or just plain letting him down! It is a rare sub who after being stunned by the onslaught will not get back on the job immediately—if the accusations are just. Some are adept, however, at reducing a patsy's state of anger, as by carefully and calmly listing why his accusations are incorrect. These same subs still will respond to the accusation—even when denying them—by doing the job immediately.

Demands are readily made in the heat of anger when, say, a sub suddenly pulls off his crew in the middle of a job, and in the middle of the day, for pressing work elsewhere. Perhaps the only way to prevent such a delay is to yell and threaten furiously—the sub cannot rightfully do it; the job must be finished! (One sub was so upset by the yelling that he brought his crew back at 5:30 p.m. to finish the job and paid them overtime.) Angry demands are also readily made when successive delays in starting or returing to the job had built up and all other calmer strategies have failed. (For example, when the electrician kept postponing the finish work, the author finally phoned and said furiously, "Well, we need you to get up here and finish, Jim, because we need to get an inspection and get our final payment!" The sub responded to the anger and acute need, "o.k., o.k. we will finish you up by tomorrow afternoon." The patsy answered, "fine" and hung up so as to catch him on the responsibility of his promise). Angry demands are also effective in breaking through the barrier that a secretary puts up to protect her evasive boss from frequent phone calls inquiring about delays. "Look, all I want to do is talk to him. We are doing business together, why can't I talk with him?" Thus, anger embodies the other strategies as it involves accusations and demands.

It is worth emphasizing that use of these strategies (of ensuring, pushing and anger) as well as their non-use, is linked with the normal tendency for the patsy to accept the natural course of things, particularly the competition of elsewhereism. He goes along with unexpectable and expectable delays when the consequences are slight, and, when not up against it, will say "Let things go at that." This stance is easier; he cannot fight all delays, and re-articulation of jobs for short interruptions is not all that difficult. The patsy, in essence, tends to conserve his energy for handling delays that have sufficient disruptive consequences for ensuing work, particularly for articulating the timing of successive jobs. We turn now to the problem of generalling the articulation of successive jobs.

VI

Articulating Successive Jobs

The continuous articulation of successive jobs is generalling at its utmost, for this is how the house gets built. The patsy's articulation of the subs' work temporally maps the course of construction—it unfolds the patsy's plans for a home; it puts him in control of the total job. Controls by subs are limited to their particular jobs, with maybe some spillover to another job on the same site. Generalling the total job is a prime source of the patsy's relative balance of equality in the power symmetry with each sub. The patsy is in charge of more than any one sub's job; thus the latter is, in some measure, subordinate to the requirements of the total job.

In his generalling the temporal articulation of many jobs, the patsy decides as the job unfolds who, when, and what type of job is to be carried out next. As things get done, new things "come up" and the patsy must then get them handled. In part, there is a natural order of work so that the total job articulates itself somewhat; but, when specific jobs should get done, and by whom, represents a continual problem of articulation.

To accomplish this generalling, the patsy has an overview of what must be done to build the house, and probably only he will ever have the total picture. All other views typically are partial to specific work and, at most, how a partial view respectively feeds out of and into the previous and successive jobs.

As we have seen, deciding which sub to employ and when he should come starts with the bidding process. The patsy starts bidding any one job according to his overview of what job will be next. As he details with the sub, together they work out the articulation of the sub's bid on timing with the total course of construction. A central aspect in choosing a sub is how well he is willing to articulate his job with others, and how well his promised time does in fact fit into the total picture. Often in bidding this temporal aspect can be broadly checked out on the phone before the sub makes a site visit. If clearly he cannot articulate his work with the course of construction because of work elsewhere, then making a bid is a waste of time. (Thus, during the initial call for a bid, many subs disqualified themselves on this ground: I am too busy" with work to meet my overall schedule.)

The typical pattern of articulation starts during the bidding. The patsy articulates the succession of jobs on his own, while both he and the sub (and often a crew) articulate the temporal fit of each job. When delays throw off the articulation of successive jobs, then the patsy and the sub use the strategies discussed in the previous chapter to bring the job back into line; meanwhile the patsy calls—in line with the total picture—other subs to rearticulate their jobs. Perforce then, the patsy is constantly generalling the articulation of several successive jobs, irrespective of whether the temporal order holds or is thrown off.

As the patsy can scarcely help learning rather early, thinking in terms of types of jobs helps him to general. These are: the *base* job, the *simultaneous* job, and the *quick* job. The *base* job is fundamental to getting many simultaneous jobs going. Once the roof is on, many "finish jobs" can start; once the foundation and subfloor are done, many "rough-in jobs" can start. Thus the base job stops all else while it is going on, and then ushers in many simultaneous jobs of rough-in or finishing nature. The base job is most affected by the contractor's elsewhereism demands, and most affects both the articulation of other jobs and the total course of construction. It also has most effect on the elsewhereism demands of other subs and the demands of the patsy. It clearly fits into the successive pattern of jobs, while *simultaneous* jobs can overlap and/or succeed one another according to the patsy's planning. Electrical, plumbing and heating can all rough-in on an overlapping basis. There are a few aspects of these jobs which clearly must succeed each other, and these get articulated on the spot. Control of base jobs constantly shifts between sub and patsy.

The *quick* job is highly necessary to get another, usually base, job going. It is something which has to be done at the last minute. A water line has to be run, a tree cut down, roof wiring must be laid before the roof, a hole cut in the floor for the fireplace, a survey stake placed, kitchen counters braced for the tile. The necessary articulation is clear. The job is needed so that another job can start, often within a few hours. When carpentry is involved, the patsy must either hustle a sub to do it or do it himself. Should he fail, he may delay the next job and lost the sub to elsewhereism for a considerable time—particularly if it is a base job. At this point, generalling can be frantic. ("Get that cat up here and smooth out the road, the paver arrives in half a day," I pleaded when a delivery truck had roughed up the base for the paver, who finally was coming.)

These jobs usually succeed themselves in progressive form. A quick job will be necessary to get a base job going, after which simultaneous jobs start until a base job is needed (sheet-rocking, for example) which in turn stops all other work until completed. More quick jobs come before base jobs, since the latter cannot be held up without causing drastic delays in many other jobs. However, few quick jobs occur before simultaneous jobs, because there is more time, offsetting immediate need, to get them done. Timing is flexible, unless a base job is coming up which necessitates fast closure to a simultaneous job. (Thus the electrician lagged behind on switch boxes which, all of a sudden, had to be mounted within hours since the rockers were coming to cover over everything permanently.)

To give the reader the flavor of generalling the total project, I shall discuss first how the patsy keeps his overview parallel with the course of construction and the articulation of each successive job. Then I shall discuss articulation of the quick, the base, and the simultaneous jobs.

The Patsy's Overview

The patsy's ability to keep his generalling overview in accord with the course of construction is reflected by his feelings about how the current and the next jobs are going. If the jobs seem "in articulation" he feels good and delighted and satisfied; if they are "out of articulation" he feels frantic and—once they are rearticulated, he feels relieved. If he feels good but the jobs are actually mal-timed or heading for trouble, reality will soon move in on him and so he discovers his mistake. He gets back in touch with the total job by hurried re-articulation, so all runs smoothly again.

Feelings are the best indicator of the patsy's overview being consonant with construction, because it is difficult to see the big picture or to describe it. As a general, the patsy is articulating successive jobs—particularly to get a base job going—and when they all integrate and fit that is a very satisfying experience. Over and over, patsies say something to the effect. "I feel good this morning, because everything is going to get done. We're going to get electricity, water, survey, and then the foundations are going to start right off the bat." Or, "You feel tremendous relief knowing that at last the cat and the pipe will be there." And, "I find it very satisfying when I can know that by the end of the day quite a bit will be done on this building site."

"The image for the total integration is in somebody's mind. But even if it isn't, or even if the whole thing seems completely out of order, somehow it all falls into place in the end and it looks all right." The notions of "falling into line" or "the job unfolds itself" indicate the faith that the patsy has in the natural order to carry him along when he loses his general view and/or things get out of order and he has yet to re-articulate jobs. Bringing the total job back into view and articulation results in a feeling of "tremendous relief." "The patsy feels a great relief to know that things are moving again." The effort, usually somewhat frantic, to rearticulate is indicated in the following statement after a patsy had been rained out. "Now we're going to get the roofer back on the job. i have to get the mason up here for the wall. I have to get the electrician back up to do one more thing on the wiring. And I have to line up the dry-wall contractor immediately." In short, many jobs—some quick—have to be rearticulated in order to get to a base job. And this view is the patsy's alone, for only he has this concern. Thus when taking stock of the rearticulation, he feels "O.K."

Being in charge, and through his knowledge of the total picture, the patsy can also provide *breathing space* for himself when he needs and wants it. By breathing space I mean that he can have the site to himself, as a respite from having subs all over the place, or can turn the house over to a base sub and disappear for a day. Bucking and handling problems caused by subs are generalling chores that are wearing for a patsy. If generalling becomes too much, he has the power to call a moratorium at any point: otherwise there are natural breathing spaces which his overview allows him to anticipate. Having the house to himself can allow the patsy to do an important job or savor a rest before the next onslaught of subs arrive and generalling problems peak again. The clearest breathing space is building the sub-floor on the foundations, when the patsy does the carpentry himself with no other subs around. This is a base job for the patsy.

In the following sections which treat the quick, base and simultaneous jobs, we shall see how the patsy's temporal overview meshes with the specific temporal view of each job as held by the sub and his crew. Essentially the sub's view of his job is 'the faster the better,' since that way he makes more money—unless it is worth delaying the patsy's job for another elsewhere. The crew's view is to fulfill the job-timing contracted for by their sub without going faster or jeopardizing quality since their *sub,* not they, benefits from the speed up. And they will "go on to another job right away," even when it makes more money for the sub. "Actually Sandy said they could finish up at noon, but really they didn't finish up 'til 4:30 PM on the nose," which was their actual deadline. When crews are not given a timing and/or are running late, the sub himself may help to finish on time because he needs the men elsewhere—and even the patsy may pitch in and help because another sub is arriving the following day. (For example, when the mason lagged a little on the brick wall and chimney, his sub came and "raked the joints" for him, saving about an hour. This relieved the mason who then finished right on time.) The patsy learns to work with the specific views of sub and crew in a fashion that fits into his articulation of the total construction.

Quick Jobs

As noted previously, the quick job is a highly necessary job that has to be done at the last minute as forerunner to another, usually a base job. There are four types of quick jobs. A *resource* job is to bring power or material to the job to be ready for the next. (For example, to start the foundation I needed a water line and electrical line run quickly. In the middle of the sheet-rocking job, the crew ran short and to avoid delaying the construction days, I had to hire a trucker to get a few sheets of rock before the crew left.) Many *services* are quick jobs, such as getting a quick survey for staking out the foundation to be sure it is right, or getting a tree removed or limbed. Others include getting a sub to install cabinets quickly before the tile setter comes the following day, or getting a sub with the equipment to come over and level the sub-floor as it is being built.

Then there are quick *preparation* jobs. They prepare the construction for the next base jobs, such as: framings for heating vents, holes in the floor for masonry, wood slats for tile counters, or rock base for paving. While they could be done early, these usually are postponed because of nuisance, or the patsy's not being aware until the last minute of exactly what is needed, or because of the hazard of leaving holes around or having rock washed off in the rain. The resolution is a last-minute preparation.

Missed jobs are aspects of jobs missed by subs, requiring that they return quickly to fix them before the work is covered over. Electricians may forget a switch box, the mason may fail to put in enough tabs to hold a masonry wall, or the plumber may forget a cleanout. Sometimes a missed job is discovered only at the last minute by an inspector. The careful patsy may have subs ready on a standby basis to make a correction quickly before the next job. This eventuality may occur with plumbing, electricity and masonry, to mention only a few of them.

The necessity to have these quick jobs done quickly leaves the patsy somewhat less protected than usual by the bidding system. The patsy may have no time to bid out these jobs with site visits. He then must bid and detail over the phone. Unless the job must be done by a sub already on the job, there is plenty of room for error in details and costs, since the premium is on time. But in the big picture, the patsy calculates when speed, to get ready for a big job, is worth the relative risk on a quick job at too high a price. The problem is not careful bidding but articulating—getting the job done quickly! The pressure is high to keep jobs moving and not to lose the subs to elsewhereism.

The patsy may use several strategies to obtain a quick job. One readily usable tactic is to *bait* the sub for the quick job with a promise to let him bid on the big job with an inside track on it. (In getting the electrical and water lines run for starting the foundation, I told the electrician and the plumber, "When you are up here, I will give you a set of plans for drawing up a bid on the electrical work for the house. And unless you are way out of line, you'll probably get the job." In both cases a man appeared within a few hours to run the lines.)

Another tactic is to *phone call* bid out a job. Quick jobs are simple enough, usually, to detail over the phone—tree removal, for example, or water or electrical lines, or surveys. Details can be stated simply and fast: "What I want done is simple enough." It can be pointed out that a short crew (one man) is all that is necessary, so that jobs elsewhere will not suffer. If the sub can barely spare a man, work details can be cut back even further to simplify the job and cut down on time. Thus the trees are cut down, but not hauled away for another week. Or the patsy runs the electrical line and the crew hooks up both ends; or the loader does not need the backhoe too; or the patsy can work with the sub during lunch hour for quick leveling; or he can pick up the bathroom sink from the plumber so that the tile job can start (since otherwise delivery would take two hours). As noted earlier, the phone call bid is chosen on the basis of promised time.

As seen above, a coexistent strategy—as well as a strategy in its own right—is for the patsy to *pitch in* and to help do things quickly. He may crew for the sub, or do the job himself, such as carpentry preparations for heating, masonry and tile contractors. For the mason, a hole is cut in the floor for the ashpit and fireplace, and in the roof for the chimney; and for the tile setter a sink, counters, and upper cabinets must be set.

When the patsy pitches in, he finds that as the general he must crew or at times be a laborer to keep the jobs moving. He notices that subs do this also for their own crews—to hasten their finishing a job. With the patsy's balance in power symmetry and because of his general purpose, he loses no power in working for awhile in the lowest job position—taking orders from crews. "All right, let's get those heating ducts in quick," as an order from a crew to the patsy is not a distressing loss of power; because the patsy has much more at stake than the crew. If the patsy slowed up the heating job, his costs could go up and delays occur in successive jobs.

While baiting on a phone call bid, the patsy may sprinkle in *referral* ploys to get a quick job. One is to mention a *referring* sub or other party, or to offer to refer the sub to friends' jobs. Thus the sub has jobs to gain or a reputation to lose. Another variant is self-referral: to call back a sub who was a previous employee or one he happens to know from some other experience. His quick response to the patsy is his way of developing a client.

The most pervasive and effective strategy to line up a quick job is insistently to explain its *immediate need.* The basic leverage is provided by the statement that another sub will do a big job which is coming right after the quick job is to be done—usually a matter of hours or the next morning. "The reason I am calling is that we just gave out the foundation contract and they want to start at 12:30 tomorrow," so that the survey was put in in the morning. Explained need is most effective because it is clear, reasonable, and within all subs' understanding. They readily feel the importance that the patsy puts on getting a quick job done so as not to hold up the next. They honor his need if at all possible.

In talking insistently, the patsy continually emphasizes his own time. 'It has to be done by noon." "We need that wire in *today!*" "It *has* to be done by the end of the week." "I will meet you on the site in an hour." This temporal insistence, with an accompanying explanation as to its need, is hard to agree to with a delay in mind. The sub either answers to the turning or declines the job because of elsewhereism. He dare not give an unfulfilled promised time and delay things, as *other subs and the total course of construction would be affected.* In insisting on temporal fulfillment, the patsy is. as a general, using his power balance in the symmetry to articulate his job. If he fails to arrange the quick job, he can look to no one else to blame when successive jobs are delayed and subs lost to elsewhereism. The only solution is again to begin re-articulation. Most subs, accustomed to such situations, respond with immediate compliance to the immediate need: "I'll be there tomorrow."

In anticipating rebuffs or delays, the patsy finds that well-prepared excuses usually are not needed. Indeed, the patsy is happily surprised and/or deflated, being left with unused justifications about why the job should get done. For example, one patsy expected the plumber to balk at setting the tub, which was needed to frame for the roof because of its possible intrusion into work elsewhere. But the plumber's response was, "I'll be there before noon today." To reaffirm the imminence of need, the patsy may use a *checking* strategy (to make sure the crew has left), or *anger* (if the sub does, in fact, start to delay the quick job).

Being conversant with the course of construction and the given type of quick job, a sub may try to overrule the need by explaining how a job can be done later, without holding up other work. The sub may offer this suggestion by saying he is just trying to save the patsy money that he would waste for extra trips or extraordinary work. (For example, the surveyor wished to wail until after dozing to put in foundation stakes so they wouldn't be disturbed and have to be reset. The plumber explained how a bath could be put in after framing.) The patsy's recourse is to be the general. Either he can articulate his way and demand the job, or he can accept the advice and re-articulate. A flimsy way for the sub to attempt overruling the need is to try to convince the patsy that the subsequent sub will not arrive on time, ". . . as they never do," so that the quick job really is not so imminent. "You can wait a day." Again the patsy, as the general, must decide whether to accept the sub's re-articulation.

Base Jobs

As noted previously, the base job is fundamental to getting simultaneous jobs going, but while it is going on all other jobs stop. The patsy virtually turns over the house, or a sector of it, to the sub and lets him keep it until he is done with his work. No other work can be done while the house is being roofed, sheet-rocked, or floored. Nor can anything be done while the foundation, underpinnings and sub-floor are being built, or excavation or painting is being done. No one can go into the bathrooms or kitchen while the tile work is going on. Once finished, as indicated earlier, many simultaneous jobs of roughing-in or finishing nature can start. Thus, once the sub-floor is done, then electrical, plumbing, heating, masonry rough-ins can start. Once the kitchen tile is finished—everyone has to stay off the tile floor for three to four days—then the electrician, plumber, heating, painter, paper-hanger and appliance man can finish up.

With the exception of the roof and driveway, each taking about a day, the other base jobs take a week or more. Their long duration allows much time and probability for elsewhereism to draw the sub away for a day or half-day, thereby delaying that job and successive jobs. The patsy's articulation problem is to get the base job started and keep it moving without delays and their consequences, while carefully warning the subs of their impending starting times. For these jobs, articulation is a shifting *joint venture in control* by the sub (with his specific view) and by the patsy (with his overview), each playing upon the other with their edge in the balance of power symmetry. The articulation problem breaks down into getting the job *started* and *managing its duration* without, or with minimal delays.

Starting: Of course the first way to get the job started is to set a promised time during bidding, but delays are endemic with base jobs since they are so frequently derailed by elsewhereism. We have seen the many strategies, by both patsy and sub, used in handling delays. Here I wish rather to discuss their joint articulation and re-articulation.

When the patsy and sub arrange a fixed promised time, it is then up to the patsy to get the quick jobs done which are necessary to begin on time. His job as the general is to articulate them so that the sub can arrive and begin on schedule. If the sub then delays his starting, the patsy has considerable leverage with which to push or with which to get angry. To eliminate a delay, he has done beforehand those things that the sub wanted for preparation, and the sub has then not met his part of the bargain. ("Fred [the debris sub] cleaned out the whole driveway. Kelly [the excavator] shaped it up a little, the grate is ready, so you can pave now and we need the driveway, WHY ARE YOU SO LATE!?" I asked the paver who had failed to show up.) Further leverage to get the job started can be applied by explaining the necessary consequences of the base job. "We need your job done so we can get a final payment, pay several subs and get the completion notice, and we need the driveway so subs can bring things to the house to finish up." In short, the patsy pushes the sub to articulate with him as planned, because of accomplishing preceding jobs and getting to subsequent ones. They then rearticulate the job, per the patsy's pressure and the sub's demands elsewhere.

To avoid the delays brought about by a fixed promised time, the sub and the patsy may decide to articulate the basic job in accordance with the course of construction. The patsy is told how to prepare with quick jobs, and the sub is kept posted as to readiness for his job. "Mr. Bridges, the tile contractor, phoned me yesterday saying that he had ordered the tile for the kitchen. He wanted to know when he could start. I (the patsy) said I didn't know. The cabinets were due any day, and we had a sub alerted to set them immediately, and we would prepare the counters per his instructions, he said to just warn him a day ahead and he could pick up the tile and start. Thus the patsy "keeps the sub posted" or alerted as to the best starting time.

In warning the sub, the patsy gives enough lead time to get materials and a crew ready. (The floor man needed two weeks notice to obtain the woods.) Also, the patsy and sub can go over preparations and articulate the base job accordingly. (For example, to start the excavator, the sub was told when the survey and tree work would be done. They then worked out getting the excavation permit and order delivery of the cement tiles, and who was to obtain both. "As soon as I get the permit I'll call you," "Fine, we'll be right down to start.") A picture is generated in both of their minds as to timing and who does what. Typically, both wish to design the articulation while doing as little work as possible to put the job together and get it started. This becomes clearer as the job progresses. (In this case, while either could get the permit and work with the tree man, the patsy did the work to ensure the articulation of the basic job. The sub sat back and waited, while keeping the tractor busy elsewhere.) Of course, as the patsy and base sub articulate, the patsy must convey timing to subsequent subs so that they will be ready to work immediately.

Duration: Once started, the articulation of base jobs is focused on handling the interruptions caused primarily by the unexpectable and by expectable delays, because of elsewhereism. The first line of effort is in preventing interruptions during the job. If they occur, the second line of effort is in joint re-articulation of the base job and possibly the quick, subsequent, successive and simultaneous jobs. These efforts culminate in keeping the *base job moving*—hence the total job is kept moving.

To keep the base job moving without interruptions, the patsy uses several strategies. The major one is *ever presence.* He must be on the job every moment to ward off elsewhereism, and to keep tab on emergent problems so that he can work out joint solutions with the sub or his crew. In so doing, the patsy may overload his jurisdiction of articulation by taking on more than his share of the articulation problems, just to make sure that the jobs will fall into line. Indeed the sub, by doing nothing even when he is supposed to, can force the patsy to articulate crews and other supporting subs—which is really the sub's responsibility in the joint articulation. Since the patsy has more at stake than the sub, he may do the work, (The author was forced to do much of the excavator's work in getting both supplies and a backhoe operator.) But what starts out for the patsy as lack of advantage in the power symmetry—he is forced—can turn, through his efforts at articulation, into his advantage. In overloading his work, he controls in his favor; this can put the sub at a disadvantage in changing or modifying the ongoing timing. The sub, however, does not lose that portion of his price for this kind of work not done, and he is free to plan and work on jobs elsewhere.

To prevent delays, the patsy can *pitch in* with the sub to check on supplies, and even run errands for minor items which have been forgotten. The patsy may even lend a hand with some of the less skilled work to assist in finishing a section of the job (e.g., help the floor man stock his job from the truck). He may even pitch in with his sub and give orders to crews or other subs (e.g., crane company) not to let elsewhereism interfere with the job as it can intrude at any moment of the day. ("Don't worry, Tom [the foundation sub] wouldn't have let him go!" was the response to my demands that the foundation excavator not leave at noon for a quick job.)

For base jobs of long duration which have intermittent different phases (sheet rock, flooring and foundations), the patsy and sub *continually articulate*, at each new juncture, the remaining timing of the job; pouring, then stripping the foundations, taping the sheet rock after hanging, or sanding new-laid floors. In this way, they work together to prevent unnecessary interruptions and make plans enabling the patsy to alert his subsequent subs.

In preventing interruptions, the joint articulation of the job is usually characterized by some conflict between patsy and sub (or his crew which is being drawn elsewhere). The interrupted job indicates the patsy has lost some edge of power symmetry. But in re-articulating an interrupted or delayed job in progress, the patsy and sub work together to get it going again.

In working toward the same goal, power symmetry is not an issue unless the patsy is striving for a faster restart time. The sub, having had his interruption, usually wishes to finish up just as rapidly as the patsy. At this point *discussion* on alternative articulation in lining up the job and its pacing is the first step into which the patsy and sub move. The discussions do not have to be exactly accurate, nor the job perceptions of either party completely in agreement (seldom do either articulate clearly or have the same mental picture) for both are pushing in the same direction—an uninterrupted conclusion to the job. If they were not, then the power symmetry would decide whose perception was correct. *Flexibility* is then the approach that each party uses, as the problem works itself out through the normal process of getting work done. This leads to revealing the work which remains to be accomplished. Flexibility allows the patsy to *wait and see* what happens as the somewhat idealized pacings get actualized. And, as the patsy watches, he is always ready with the above named strategies to prevent new interruptions, primarily by his *ever presence.* If things get bumpy or out of articulation, the patsy can defer with trust to the sub to gel his crew back on track because of the consonant goals of both. The sub takes over and the patsy sits back. ("I trust you, Gordon, you do what you feel is best," was my remark to a sub who had promised that no further jobs elsewhere would interrupt his work for me.)

Given the consonant goals of getting the job done, the lack of need for the use of power symmetry, the articulation and working together, and the promise of finishing a base job which will usher in a number of simultaneous and successive jobs (which occur at this juncture of no interruptions), patsy, sub and crews all experience a period of elation as the construction moves forward. Generalling is easy and fun, since keeping the job moving is nonproblematic. "Well, great, everything is working out all right," remarks the patsy to the sub as both rearticulate the conclusion to the house's foundation.

Simultaneous Jobs

Once a basic job is completed, it ushers in several simultaneous jobs in the same area of the construction. These can successively overlap or succeed one another. The patsy's job is to articulate the initial and subsequent arrival of these subs, and to articulate their working together without clashes on the premises. Generalling at this juncture is at its peak for the patsy: he is handling a number of subs at once, and successively working with each while keeping all working in some articulation toward his overview of what should be achieved in toto through the accomplishment of separate, integrated jobs. He alone is in charge at this point; he is responsible and concerned for the overall project. The others are busy with achieving their specific jobs.

For example, once the sheet rock is done, all the finish jobs can start, such as painting, tiles, electrical work, plumbing, heating, doors, glazing, trimming, formica, cabinets. It is up to the patsy to general these many subs through articulating their work toward the construction goal, and specifically toward the next basic job, the flooring. He becomes a maestro!

Generalling the simultaneous jobs during rough-in and finish phases of construction is easier than generalling the quick or base jobs, since elsewhereism has less effect in delaying these and subsequent jobs. Timing is more flexible since any one job does not depend on another. If one job is delayed by jobs elsewhere, work on another job can go on. The patsy then can be less pushy in his strategies to bring the delaying sub back to his job. However, flexibility ceases and the patsy must push for completion (as with quick and base jobs) when a number of simultaneous jobs are drawing to a close but one sub is still drawn elsewhere, posing threatening delays to the next base job. For example, whether the plumber takes two or four days to finish the fixtures usually will not matter, unless he is the only one holding up the linoleum contractor (a base job).

Generalling at its peak is compounded by the typical pattern of subs running simultaneous jobs on a *comeback* basis. While they might have to start on time, since the time that they finish a job is so flexible, its duration need not be accomplished on a straight-through basis. A few hours here and there can easily finish a job on time, plus that these subs are experiencing the same flexibility on jobs elsewhere. Thus, they can hold down many more jobs on a comeback basis than would be possible on a straight-through basis; but the patsy has negotiated the job on the latter basis with the idea of watching construction so not to cause real delays. Until and unless the patsy learns to expect a continual moving in and out of crews and subs *at any time,* he experiences a constant sense of delay and employs the tactics discussed in the previous chapter. Once he understands the pattern, however, he generals the subs in terms of it; therefore articulates comeback times into the total picture of construction. He begins juggling his subs with comeback negotiations, much as they are juggling their clients. The patsy applies his strategies with flexible ease in consonance with the flux of the ongoing situation, knowing that "things fall into line" through his efforts; knowing also that those of the sub will too, as he gauges the course of construction and checks new temporal moves with the patsy. This becomes the normal course of generalling simultaneous jobs.

Once the patsy has accepted the comeback basis of simultaneous jobs, *the basic strategies of generalling* become: *"let the subs gauge"* the course of construction; while he remains ever ready to engage in daily, *on-the-spot re-articulations* between jobs, then to follow through when necessary with *nursing jobs along,* with *warnings,* and *comeback negotiations,* and with *demands.* These strategies fit smoothly into the active, moving nature of the construction brought on by several crews working with dispatch on the premises.

Letting the subs gauge the course of construction poses the problem of getting them started and coming back in accordance with the patsy's overview and not with the sub's specific view and convenience. For example, the patsy may feel he can move the project along faster by getting a simultaneous job going without a base job being finished—such as getting the rough-ins on electricity, heating and plumbing before the roof deck is completed. Or he may want the job started to get a loan payment. This does not usually jibe with the sub's notion that to assure the crew's getting in a full day's work, it is safer to have the roof deck completed. Subs like the carpentry preparations completed before they arrive. Their first strategy is to demand its completion, and if necessary explain how to prepare it—such tasks, for example, as rough-ins for heating ducts. If the patsy presses to get a sub started before, in the sub's view, the construction warrants, the patsy is likely to fail. The sub may just ignore the request and reiterate: "Call when the roof deck is on." If the patsy presses with, "We have half the roof deck on," the sub may talk past him or double talk, with assertions that the job will be done in time. If the patsy still presses, using the pushing strategy, the sub who wishes to maintain good-will will explain the necessity that everything be ready, thus eliminating unnecessary trips. Of course, this belies the comeback nature of these jobs, which reveals that the number of trips is dependent on many other considerations, particularly on demands from work elsewhere. The patsy's only recourse to shift the balance of power symmetry to himself is to become angry and threaten the contract: to which the sub may give token acquiescence with the promise to send a man up to get started "when he's free." When the man does show up, it is to keep the patsy happy and hook him into the job. The sub just does not start, or "really" start, until the course of construction in his view warrants it. However, the patsy's generalling gains edge in the power symmetry.

(In one instance a man was sent up, only to find that I had misjudged and the simultaneous job could not begin! The wasted trip taught me to trust the subs in their gauging of this condition and to general from there on a comeback basis. I never did regain the trust of one sub who had believed my inaccurate rendition of the course of construction. Subsequently the sub always grilled me on construction stages, then inspected, before returning his crew to the job. For example, he checked to ascertain that all blocking was in the framing before running the wires. If, he explained, one block was missing, it could mean another trip and could demoralize the crew by preventing closure. My ability to general this sub in relation to other subs had been completely undermined by my error.)

Once the construction is flowing, simultaneous jobs are *re-articulated on the spot* as the work of one sub affects the work of others, hence changing the picture. These are flexible, readily apparent negotiations, usually accomplished with the crew. (For example, it was evident to the plumbing crews that "We'd better wait until the furnace is in place before we place the water heater since the former can force the moving of the latter." So after finding out the approximate day that the heater was coming, the plumbing crew planned to return the day after with the water heater and a comeback was articulated. Before leaving, however, the plumbing crew requested to be told if the heating crew was detained "elsewhere" and did not place the heater on schedule, indicating awareness of this pervasive condition. They then closed by articulating when they could lay the water and gas lines "in time to catch up with the heating contractor.") The patsy's generalling at this point is to okay the articulation in the service of the larger view.

Most re-articulation revolves around keeping simultaneous jobs successively paced in relation to one another, such as the previous example illustrates, or as is illustrated by the plumber who planned to install finish fixtures as soon as tiles and formica were set. Generalling becomes a mastery of natural flow, and when jobs go a little awry, are slightly delayed or mistakes are made, rearticulations take on a "nursing along" appearance. For example, a door is wrong, so it is sent back and trimming and painting are done in other areas until in a few days a replacement door arrives. Complex jobs (such as doors) easily generate such nursing along, as it is hard to get the job straight at the first try. In nursing along the construction, the patsy can let a sub be delayed elsewhere for a few days, because of the existence of problems that have no undue affect on construction. Simultaneous subs do other work while waiting, or come back "most any time" according to the patsy's re-articulations of connected aspects of the simultaneous work.

The subs ask only to be warned, for they are flexible. And even if the patsy must push a little to get a delayed sub to return, it is not too much pressure for the sub since elsewhereism demands for simultaneous jobs are flexible too. This type sub can keep most clients happy since, when pressure mounts, there are always a few hours in the day that can be taken from another job.

In taking off before closure, in these situations and in general, crews are always careful to explain that they will return as soon as needed, or as possible, to work on the jobs and will verify that they are not holding up the total job by leaving. "We haven't been holding you up, have we?" queries the crew, intimating that the patsy can trust their return when appropriate, with no delays.

On-the-spot re-articulations are required of crews, since the sub may be miles away. Thus crews run their own jobs at this level—which gives the patsy an edge in the power symmetry in making articulations stick. Crews find it harder than subs to disagree or to say no. The patsy nurses the crews along by telling them what a good job they are doing, and by using the responsibility strategy of expecting a comeback at a certain time. Seldom are such commitments by crews overruled by the sub.

Articulating simultaneous jobs can become more difficult under certain conditions which are out of normal flow with the course of construction. One such, mentioned earlier, is that of trying to get a sub to start before a base job is done, such as getting the door sub to deliver doors before the sheet rock is pasted. Another condition is when it is unclear which sub must do a specific task, such as: hook up the gas line to the furnace (the heating or plumbing contractor?), provide or install roof jack (the roofer or plumber or heating and sheetmetal contractor?), or flash the chimney (the heating contractor or roofer?). Then the patsy must tease out the proper jurisdiction and get the job done. "What was in first, the gas line or the furnace?" may decide who should have attached the two—that task falling to the last sub on the job. This, then, determines who must return so that inspection can take place within the day. Whichever one returns, the patsy must then handle his anger at the other subs, as it relates to keeping the job moving, while making stick the demand to get the job done. Again, knowledge that these subs always have a crew that can be freed from elsewhere for a few hours makes the demand for proper liming less difficult.

"Loan payments" is another frequent condition which generates re-articulation of simultaneous jobs out of their normal flow. The loan company designs its payment schedule to force job completions as quickly as possible. Consequently, the total construction does not lag. This can force the patsy to ask a sub to do a little work before it is called for in the course of construction. (The incident with the electrician who wanted a finished roof deck, resulted from the request to put a little wiring in the walls so that a loan payment could be legitimated.) Mostly, though, the patsy must hurry up subs who lag, thereby delaying loan payments. His best bait then, for re-articulating jobs out of normal flow, is to offer quick payment for the job—something which all subs need.

When re-articulation of simultaneous subs becomes difficult—because one sub's excessive delay has caused the further delay of other subs, jamming the normal construction flow—the patsy uses the strategies discussed in the previous chapter to get the delaying subs going. But at this juncture many subs are involved—especially when loan payment is in the offing but it too is delayed. Then the patsy is offered much advice as to handling the delay from other subs affected who, by the nature of their interconnecting work, know several ways to keep each other moving. However well-intentioned and supportive this abetting of the patsy's generalling, it is usually to little avail since the strategies are not usually imbued with the patsy's larger view of the construction. The suggested strategies revolve around demands for a new promised time, or being cancelled, or being understanding and patient. They lack the gradualism and varied subtly gained only by being in the position of generalling several subs simultaneously, making referrals, hiring for other jobs, making private deals in playing for the larger stakes of construction, and so forth. Thus, even with this advice as a potential inroad over his power—and even while being caught in the interconnected maze of simultaneous subs—the patsy remains ever the general. He has no choice if his project is to proceed.

If, in articulating any job and successive jobs the patsy does not insist, push or pressure *every day* to some extent—whether flexibly or sternly, whether for quick, base or simultaneous jobs—the construction will glide to a premature halt amidst a mass of errors in temporal articulation and quality.

To continue and round out our development of generalling, let us now turn to the major problem of generalling, that of quality control.

VII

Quality Control of a Job

Generalling the quality of a job revolves around getting the details straight and preventing errors. For the patsy this is not an easy task, because he must inspect, evaluate and administer the work of an expert—which experts do not enjoy from someone less knowledgeable. The expert wants to please the patsy on his (the expert's) terms, not on the patsy's, since he is the expert—and should only be judged by one! The patsy then, as general, starts out with an edge in the power symmetry over quality issues; but is constantly being undermined by the experts who think they know better how to do things. And knowing better they often do things on their own without asking the patsy. This, often as not, leads to *errors* of a structural nature regarding overall planning and errors in matters of taste. The former occurs because a specific view about convenience (e.g., flashing the skylights), and the latter because the sub redesigns in terms of standard ways of solving minute problems of construction (e.g., the height of the hearth.)

Therefore, to general jobs properly, the patsy must be alert at all times to watch for and hedge against impending errors; otherwise he can be sure that things will go wrong. And he must correct errors immediately, hard as it may be for him to confront the expert crew or sub, or the error "will chase him" for the rest of the job. He will always run into it, as it compounds additional job corrections; indeed, he may have to live with the error for the rest of his life. Classic examples are a doorway which is two inches too narrow, and a misplaced skylight or floor tiles not properly patterned to the patsy's taste. These small errors are legion in number and, as they are literally built into the job, often highly visible unless corrected.

The patsy's hedging against errors begins in redetailing the start of a job and continues while running it as the supervising general. In the first section of this chapter, I shall first discuss starting the job from the view of how the patsy hedges against errors as a prevention strategy; then I discuss running the job, which entails constant watching of the course of a job in order to continue hedging against errors and to correct those that occur before they have negative consequences. Last, I discuss the touchy problem of a patsy inspecting and administering an expert's work.

Starting the Job: Hedging Errors

Redetailing the starting of a job is vital because it often solves an ironical problem which emerges at this juncture: *a communication gap.* This problem occurs because seldom does the detailing which occurred during on-site bidding—a week or weeks before—reach the job intact, if it reaches the job at all! The problem is, of course, especially prominent with phonecall or plan bidding. Thus the patsy, to get the job started must redetail to forgetful subs the precise matters which will tailor their job to his construction. More important, the patsy often must redetail to a crew which knows little or nothing of the details of the bid—the sub never told them anything, knowing the patsy would detail them. Or, if the sub did detail the crew, then distortions through forgetfulness, rushing, unique imagery, and/or not very bright crews, lead to the patsy's having to set the crew straight on many matters. At this juncture the patsy realizes that detailing for a bid is *literally* so. The carry-over to the job of precise bidding details is tacitly his task as the general in the eyes of most subs. That is paradoxical considering that the sub probably was chosen because his grasp of the bidding details had instilled trust.

For complex jobs of some duration (plumbing or heating rough-in), the sub may accompany his crew to the building site and detail the start of the job. Or, as in one case, he may send his crew with a detailed map of the construction which showed how the job was to be done. But even in these instances of clear detailing, *the crew still turns to the patsy for a redetailing.* (For example, the plumbing crew spent an hour redetailing the job to me after the sub left. This was to confirm the sub's imagery, and to see if any new ideas or problems had emerged which would effect changes.) Such transactions provide a grounding of the relationship that will endure for several days during which the patsy as general will inspect, evaluate, and administer the crew's job. Such an edge in power symmetry requires clear and mutual understandings from the start.

A few exceptions on my construction were excavations, tiles, and chimney masonry when the crew's imagery, received from the sub, had to be left untampered with so as not to boggle them while they worked out the emergent, on-site solutions to the job. Then the patsy must learn both "not to butt in"—*generalling discretion*—and at what stage to step in quickly to prevent or correct errors. Thus, to general well, the patsy has to learn to pace with work: when to exert his power usefully, and when to say "nothing" so as not to interfere. In starting, he has to give crews of these jobs the benefit of the doubt until they clearly indicate an insufficient grasp of the imagery of the job, or that their judgment can clearly be trusted. This is a risk the patsy must sometimes take to get the best job. It is difficult to test for the crew's imagery, since they cannot verbalize well what they still can know very well by "know-how" and experience. Generalling discretion requires that the patsy not interpret non-communication as an ordinary communications gap.

Reducing a genuine communications gap by redetailing amounts to lowering the probability of errors. The chance of errors varies with the degree of carry-over from the bid. Most errors come when the crew has been told nothing, and the least when the crew has had some briefing. But redetailing only reduces errors, it never quite results in conveying the exact imagery of the job. It only brings the job as close to the desired end as possible, which is usually well within the range of expected quality.

Redetailing, of course, puts the patsy in complete generalling authority. A crew may come fully unaware of details and have just a general perspective, i.e., rough-in wiring. The crew cannot be assumed to have been briefed until proven so. If a crew walks on the site and asks, "Where's the funny papers?" (plans) it means they need detailing. If not, the patsy is in independent control over them as to details, with no interference from the sub. (This independence in one instance—electrical—persisted each day to a new and different crew which was sent up to go on with the job. And the new crews did not get briefed by the previous crew. This turnover required the author each day to start with a crew to be sure the job was done right. He had to be on the job every morning early; his generalling capacity was obviously being used by the subs.)

By-passing the sub and redetailing *directly* to the crew has several benefits for the patsy. It is easier to give the crew details first hand than work through a sub, who may be on the job only a few minutes a day. Having to work through the sub may cause delays because one must wait for him to show up again on the site to relate a problem to the crew and may cause errors by asking him to give orders on a job he is not as attuned to as the patsy and the crew. One less person leads to less distortion. Dealing directly with the crew also allows the patsy an immediate response, authority over the specific person who will get the job going, and a constant ability to troubleshoot within moments to prevent an accumulation of errors. Dealing directly also allows the patsy to shepherd through a slightly illegal detail for which the sub will be responsible during the inspection—for example, spacing metal tabs on a brick wall to hold up studs. These slight transgressions can be done behind the sub's back. Dealing through the sub hampers this flexibility.

Subs who are aware of the convenience to the patsy of generalling the crew directly, may take pains to assure him of the crew's competence to do the work, grasp the imagery and work with him. Some crews who feel too good at their jobs try, however, to lord it over the patsy since they are the experts. Their deference, then, is condescending, and they are not attentive to the patsy's ideas. Then, the sub must intercede to be sure that the crew pays attention to the patsy's orders.

These consequences of direct generalling of the men who will do the job apply also to subs who do their own work, but with exceptions. A sub may forget and distort, but redetailing soon refreshes his memory of previous work. He can then, better than a crew, tailor his resources of time, money and skill and his responsibility to the job, Since he knows better what is involved for his business. But he too is more liable to the intrusions of elsewhereism which, at this point, may delay starting the job or perhaps prevent his grasp and understanding of it because he is working as the responsible party on so many other jobs. His mind can get spread too thin by elsewhereism demands and thus a communications gap is generated. Since he is the responsible party, however, if the rede tailing gets confused it can be straightened out at the sub's expense. Redetailing directly to the sub makes the edge in power symmetry more of a toss-up than with a crew, who are clearly of lower status and defer to the patsy within the limits set by their boss.

Redetailing to reduce the communications gap between the bidding and starting of the job leads to further strategies for hedging against errors. These include: providing preparation by detailed instructions, checking materials and details for quality, and adding new details more suited to the quality of the job.

We have seen previously in the discussion about generalling a temporal control of jobs, how crucial is preparation of a job for preventing delays and making the sub responsible for starting on time. This temporal preparation is set during the redetailing for the job; at the same time, the specific details of how to prepare are gone over to ensure the quality of the job. (Examples are: the frames for heating vents and fireplace hearths, the leveling of foundations, the slatting of kitchen counters for cement, the preparation of the roof deck and skylights). These preparations are the foundations of quality and must be suited to the subs' styles. When preparations are not exact enough the subs complain and disclaim potentially "lousy" work on their own part. ("These foundations are not level and our contract provides for level foundations," the mason complained. The patsy then prepared the foundations with additional cement to level them, protecting the quality of the fireplace, hence maintaining his edge in the power symmetry to demand good work of the mason.")

In redetailing the job, the situation presents itself that the patsy can *check out* the quality of the materials to be used. Variations easily occur between bidding and purchase time, as supplies run out or new materials emerge. (For example, the roofer wanted to try a new insulation material that had just been marketed; the door man wanted to make the jambs out of cedar instead of redwood, because that week it was cheaper.) On the detail level, the potential variations that affect quality are fantastic. Often what is bought is not exactly like the sample. The only hedge for the patsy is to o.k. the material on the spot, such as the grain of mahogany for the cabinets or for the hardwood floors. (The extreme of this checking on materials occurred with the tile contractor. He came for the initial detailing with a large box of tile showing variations in colors, styles and trims. He presented a tyranny of alternatives. His risk for putting in a tile that later I might decide was not right and then demand its removal would be too costly. So his protection was my agreement beforehand. But that gave me much edge in the power symmetry to hedge error.) In other contracts, previous agreements might not be so important—e.g., the new type roof insulation—so the patsy must demand this checking of materials before the sub installs whatever he is currently using. (For example, the plumber put in bathroom basin fixtures without asking, before I noticed, and then I had to pay extra to have those of better quality installed.)

Not all checking is on materials. Redetailing reduces the communications gap and at the same time allows a checking out on details of the bid. This is particularly important for complicated bids, which vary from those for standard jobs. (For example, the door head lines were lowered two inches in the main part of the house to stay on the same level as the window lines, which detail had to be checked out for each door. Door swing openings, size of jambs, bronze or brass locks had to be verified also. Roofing skylight openings varied from standard so had to be re-checked.) This final checking of material and construction details clearly puts the patsy in a generalling posture to hedge errors, since he is committing himself to a quality of job for which he will pay. Once finally committed, his only recourse is close watching of the ongoing job to be sure that what is happening was what was detailed for.

The redetailing and the checking of details easily lead to new details which will enhance the quality of the job. This possibility raises two issues which the patsy must general carefully: extra costs, and matters of taste. New details may change the bid, because they require more or less time and/or more or less costs for different or additional materials. The chances are that the patsy will get no credit, unless it is completely obvious that credit should be forthcoming, and that the sub will try to charge for extras. This means the patsy must do some quick checking of prices and curbing of the excesses of the sub.

Affecting these extras also can be affirmations of taste by the sub and which do not please the patsy. The latter can be told that the extra cost is worth paying because of a better job or because of avoiding a poorer job. Virtually every sub has his bag of small but extra details to add to the job. The patsy must general firmly over what to do in order to maintain or enhance quality through new details—or he loses his edge in power symmetry. Since the sub's taste usually or often is not that of the patsy, the difference is strengthening for the patsy's edge in power symmetry.

If the new detail is associated with technical considerations, then it is hard for the patsy to refuse it. It may be a new style of cleanout on waste pipes, a new kind of roof jack, a different and more serviceable formica trim, or a different grouting of the fireplace. At this point, the patsy may need outside consultations to withstand the sub's new advice.

One way of obtaining the extras without cost is to ask the crew to do it or accept the crew's offer, say, for a pressure reducer; then the patsy says no more and leaves the outcome to the crew's ability to remember to advise his boss. Sometimes the crew just puts in the extra as a favor or to maintain a compatible relationship with the patsy. The crew's loyalty to the sub is never too strong.

Redetailing and checking details can also, in leading to new details, start anew the *redesigning process.* For example, the finish jobs (paints, tiles) which admit of so many alternatives, easily get redesigned. That is almost inevitable. Rough-in jobs are redesigned at the start in only minor ways, if at all. Thus in hedging against errors, these strategies if not generalled properly open up avenues for even more errors if new details, and especially redesigned ones, are allowed to get out of hand. This can occur when the bidded imagery is changed; then, room for errors of taste through style differences may easily slip in unless everyone is alerted. For example, if the patsy redesigns with one crew, and a new crew comes the next day with the sub's imagery before the patsy arrives, they may work an hour on wrong details. The patsy, therefore, briefs everyone on all redesigning—at times even including the secretary back in the office who may coach a crew in the absence of the boss or foreman—since otherwise everyone starts using his own judgments of taste to keep the job moving which, in the patsy's view, can be mediocre or routine. In generalling everybody, at the start and even more so during the job, the patsy changes and redesigns imagery with caution. Changes can become a disaster unless closely watched. For example, unless clearly laid out, variations on the trimming of tile, or the heading of a fireplace can vary fantastically by the time those jobs are completed. Vagueness or changes may put a crew on its own, and this can make for taste errors which are very hard to correct. By the crew, the errors are excused, explained away, or rationalized as just as tasteful ("floor tile also may vary by ones not two")' The patsy then must live with the errors until later he reconstructs his house. The sub, on his side, has his contract to back up his work commitment, in which taste and often technique are not at issue. The patsy must get his protection against redesigning by his own generalling firmly and carefully. Changes are then picked up quickly. When he does not check carefully, changes are *discovered*—but often too late.

Of course, redesigning to change or improve quality of a job has a temporal dimension. First, it provides an opener to a sub who likes to redesign, and this takes time for the patsy to handle. The time is taken because the details concluded at the bid now are not viable and have to be tediously worked out again. Second, the redesigned job may take longer, as well as cost more, because of lack of materials or extra work involved. This may throw off the construction schedule and delay other subs, if it is a base job. Thus, at such a late hour, redesigning is safest with simultaneous jobs. If this redesigning delays other subs, it can generate rush jobs for them in maintaining their schedules elsewhere. This could hinder the quality of the jobs done for the patsy. While generalling his way out of this bind the patsy will be told by the sub that they "had to rush" because he altered the schedules.

Running The Job: Close Watching

Several sources of potential errors or poor quality bear continual watching while the patsy is generalling the jobs. A partial list includes: level of workmanship, incompatibilities between patsy and crew or sub, incomplete or premature closure, accidents, rushed timing for the job by the sub because of elsewhereism or because he is running out of money, cheating, incidental redesigning by crews during the jobs, by-product damages to a job caused by crews doing other jobs, and differential judgments between patsy and crew or sub on what to do next.

Accidents, cheating and premature closure speak for themselves and are hard to control. Corrections must be demanded on the basis of guarantees or morality. By-product damages refer to, say, a plumber who damages wiring while installing pipe, or a painter or dry wall crew who smudge wood trim that they should avoid. Fortunately, most crews are very careful of other work. The patsy can warn crews who appear careless or chastise crews who are careless. These problems are fairly straightforward and when they cannot be controlled, they are quickly caught by close watching and then corrected. Timing and elsewhereism problems which occasion too quick a job, and undermine quality, we have discussed in previous chapters.

In this section I shall discuss briefly first, the level of workmanship, then the main source of errors of quality, constant differential judgments of what should be done, and finally incompatibilities.

Level of Workmanship: The patsy has little pre-control over the level of workmanship once he has made his bidding choice, which presumably considered this aspect of work. He now deals with a range of acceptable work and by close watching will spot, as immediately as possible, work outside this range. The patsy can develop a *probability notion* based on status: sub, crews, apprentices, and labor or an "extra man" whose work varies downward in quality and must therefore be watched even more closely. (The floor man was drawn elsewhere by heavy demands with his whole crew, so hired an extra man to finish the baseboard. The extra appeared to be all thumbs and I watched every move to keep his work up to acceptable standards.

But this probability notion—that the higher the status the better the work—can be inaccurate, as pointed out by a plumber sub: "Actually, you know, apprentices sometimes do the best work." "Why is that?" "Well, you know, these apprentices work four years trying to get their plumber's license and they have to do a good job and they work hard.

They try to keep perfect and they work fast so you can depend upon them usually to be better than the guy who has his license and can start relaxing." But he was referring to apprentices who had enough experience to develop skill.

Another probability notion is based on "danger hours." Some crews constantly bungle (doing wrong cuts and so forth), but a pattern of slight bungles emerges among all workers after three in the afternoon. The pattern is based on fatigue and affects the most skillful. It is common to hear crews say, "We'd better start quitting at 3:30 or 4:00 PM before we hurt ourselves." When the patsy sees them tiring his only recourse is to ask the crew to leave, hold off certain jobs or slow down, and come back in the morning to work on the difficult aspects when they are fresh, accurate and cheerful. If not too fatigued, the patsy needs to increase his own vigilance when the danger hours approach.

At the other end of the continuum are crews who display high skill, industriousness and conscientiousness. They work to perfection, take no short cuts, and stick to a job until it is right no matter how much time it takes—even when it overruns provisions of the contract. These highly trained crews have come up through an apprentice system which is linked with a trade union. Examples are the plumbing, tile, masonry or rocking crews. Once the patsy has spotted such workers, close watching can be decreased to intermittent monitoring; although as we shall see in discussing differential judgments below, problems of taste can never be left to crews. The most conscientious crews are actually best at routinizing interesting variations.

Differential Judgments: The cutting point for the patsy's generalling ability in keeping the edge in power symmetry over quality control with the sub, is his handling of differential judgments about how the job is and should be going. To tell an expert that things are not going right is not easy for a layman, but the patsy has the initial power, right and pressure, as general articulator to do so. Also, as an owner who will live with all the results for years to come, he is personally impelled to speak up. Thus, positional and personal concerns are two sources that make the patsy's judgments of the ongoing different from the sub or his crews.

Other sources are the emergent and constantly ongoing quality of the job which presents a series of occasions for quick, minor decisions. Left to their own, crew or subs will make their own decisions (level of wall sockets, tile patterns) which will have implicit redesigning in them and lead to routinization of the patsy's interesting variations.

The greatest odds are that no matter how trustworthy in skill and taste is the crew or sub, their choices will be different than the patsy's since usually alternative choices are many. And the odds are even greater for taste decisions as compared to technical decisions, since experts hold greater sway in the latter and are more pedestrian in their handling of the former. Also the more complex, the less clear, the more evolving are the jobs, the greater the chance of disparate judgments.

The cardinal rule for defense against mistakes and control of differences is close watching on the job *all the time:* a rule to be varied at the patsy's discretion and risk. (For example, arriving only fifteen minutes after tree work had begun still resulted in losing one good tree.) If the patsy is not there to answer questions and stop mistakes, the crews will continue. (Only one sub made sure I was briefed at every minor juncture; the rest would never delay the job for, in their view, minor variance decisions.) And as I have said, short of new construction, many decisions made by crews are irrevocable. As one general contractor said, "You have to be there every minute or something goes wrong or work stops." Even general contractors have the same problem, though less so with subs whom they employ over and over. The extent of watching (but never loss of immediate contact potential) is less necessary, the more simple and clear the job and the more skilled and conscientious the crew. Yet, on all evolving jobs the patsy may have to sit, wait and watch patiently, because leaving for a moment invites "disaster,"—as in excavating a difficult lot for a building site or in masonry work not routinized. This is the only way to be sure of exact quality. Most jobs require intermittent monitoring along with instant contact potential—which means the patsy is working somewhere on the site and his work is open to constant interruptions.

This condition of handling differential judgments, resulting mistakes and varied unwanted taste consequences, generates a motivation to watch the jobs closely—beyond what generalling still generates—called "patsy worry." You cannot tell what will go wrong until it hits and then it may be too late—thus, keep watching. The patsy's power to hire subs and to give them orders to get work done is a quite thrilling power at times—"move that hillside" he can say to an excavator. *But,* the obverse is the powerlessness felt when an error is too late to correct, as when a crew or sub takes off on their own divergent judgment. That is a powerlessness born of loss of control.

In practicing the close watching to prevent the consequences of differential judgments, the patsy then puts into service (for generalling) several, subsequent, resolution strategies. These resolve the differences and possible errors. They also are part of generalling the subs and keeping the edge in power symmetry. The strategies are needed since close watching in preventing subs or crews from following their own judgments will generate the demand for a solution to replace their denied judgments. A follow-through is required to keep the job moving.

As a general, the patsy can give orders to a sub or his crew for that is within his jurisdiction, but as a patsy he is a "know-nothing." The sub knows more technically, but the patsy knows what he wants—taste. Thus, most of the patsy's resolving strategies favor taste issues in order to keep the patsy's control, since taste as owner-builder-general is more within the patsy's jurisdiction than techniques are. At this point, the patsy must have the temperament referred to in Chapter I in order to act strong enough to use and keep his edge in the power symmetry.

The patsy uses three kinds of tactics depending on whether he wishes his judgment to *prevail*, be *joint*, or to *defer* to the judgment of the sub or crew. One source of prevailing where no strategy is needed is to be granted control by the sub who is leaving for elsewhere: "Now keep track of the job, be a good contractor." At the other end of the continuum are "stand-pat" strategies by which the patsy prevails at all or most costs. In the middle are pacing strategies, where judgments are provided at the right moment—through close watching—to maximize their chance of prevailing.

Prevailing: Standing Pat in the face of a counter judgment puts pressure on the patsy and crew which can constrain work. To prevail the patsy may deflect the pressure so as to reduce it. He may joke about his stand to depressurize the situation. ("Well Bill, we're leaving you an annuity in those skylights, when they start leaking you can come up and fix them," was the joking stand to the roofer's disapproval of the skylight details.) Or he may explain his stand to a crew which is literally blocked by a differential judgment, so that his position makes sense, such as: why heating ducts are underneath the doors rather than the windows. Giving a little or backing down as a compromise without compromising the stand, is an effort to prevent the patsy's firm position and the crew of sub's expertise from coming into direct confrontation, while still allowing the patsy to prevail. Acknowledging expertise but asking for the owner's prerogatives, also softens the pressure.

Another way to reduce pressure is to inundate the situation with a demand that the direction of work be corrected: "Say, we are in no rush. Slow down and do it right." Or the patsy may, upon inspection, refuse poor quality work and demand that it be done over properly. These demands are made in a "non-budge" tone. The patsy deflects or demands, using the best strategy to keep the job moving, and being mindful that too strong a stand at times can bring a halt to work or to the best quality that the crew can put out. Taking too much edge in power symmetry undermines crews' and subs' motivations and satisfactions in work.

Perhaps the patsy's best strategy for prevailing, yet respecting those motivations, is pacing his judgment with both the course of work and the mood of the crew. Pacing is a combination of saying nothing, close watching, and offering a stand at the right moment on how the job should be done. It is a difficult strategy for the patsy to apply because of his relative inexperience, the difficulty of saying nothing and waiting quietly when he would rather scream, and figuring out how to express his stand so that it fits rather than interrupts. Waiting can undermine the patsy's motivation to prevail, since it calms him down when what he needs to speak up is the strength derived from alarm or anger. When offering his judgment, the patsy keeps it as simple as possible so as not to confuse the crew, hence affect the job adversely, thus changing the pacing too. He makes simple but firm suggestions which are to be taken as orders; or he progressively details the job as it unfolds, one detail at a time—which clearly requires ever present watching; or as the job progresses, he clarifies the details of his stand just before the crew starts each phase.

Joint Judgment: While deflecting, demanding or pacing are used by the patsy to prevail, these strategies can easily be transformed into the first stage of joint judgments if the crew or sub strongly and/or cogently enough counter with their expertise. Stands are also made jointly, hopefully, when the crew checks out routinely the next move with the patsy to be sure it is O.K. This checking out is virtually an irrefusable invitation for a joint judgment, unlike prevailing strategies which require only on-site presence and intermittent watching for quick access and decisions. Once established, a crew naturally seeks joint judgments with the patsy, in favor of going on their own. That becomes the mode of operation. As a general, the patsy seeks to establish this mode whenever and as soon as possible with each crew: it is his favored protection against the undesirable directions in which subs and crews take their work when unsupervised. It entails also the least pressure on both patsy and crew which is normally generated by differences and close watching.

Deferring to Sub: The patsy can virtually avoid all strain and pressure from differential judgments by deferring to the sub's or crew's judgment. This deferring need not entail a loss of edge in power symmetry, if done as a sensible move by a general in deference to an expert, in terms of the former's power to approve and implement the judgment. This is an especially important strategy in evolving jobs which have no clear imagery (such as excavating). The patsy defers and depends in large measure on the judgment of the crew. But, as the job clarifies itself toward its completion, then the patsy switches to pacing his own judgments and joint judgments, so that in the end the job becomes what the patsy desires. This is also an important strategy to use with crews who listen but still do jobs in their own way, if the patsy recognizes that their long experience and high skill warrant it in the face of his own naivete. Not to defer would jeopardize his edge in the power symmetry. He would be seen as too naive to general effectively.

At times when a job goes astray or is too unclear, both patsy and crew may defer to each other to avoid risk and responsibility. A joint non-judgment is then reached. The resolution entails forcing the patsy to take a risk on a direction that neither party is sure of; as general it is his responsibility.

Deferring on the patsy's part may be either a result and/or a source of "bringing out" the instructor in a crew or sub. As we saw in discussing detailing bids, a sub or crew may like to teach and the ongoing job provides an even more excellent terrain for the patsy's training. Now the gist of instruction can be action, not words. In deferring to the instruction, the patsy puts himself in the role of the student—which has several consequences for both himself and the sub or crew. Primarily it places the patsy in the role of "know-nothing," and he loses his edge in the power symmetry on the job. But built into the student role is a resolution for regaining the edge of control needed to general the job: a patsy can be a student only so long without learning those details which, combined with his responsibility as general, again puts him in control. In the instructor's eyes, he has to learn, and it is likely that his subsequent control will be compatible to the sub.

But during his "training period," when he loses edge to the crew or sub, much can happen to the ongoing job. The sub qua instructor takes over as almost the *sole judge* of quality on the technical aspects of the job. He judges material, workmanship, timing and techniques. As a student, the patsy has little ground for rejecting anything in the face of explanations and details of instructions! The patsy reasons that as the job emerges he will understand better how to fit it all together in terms of the techniques that he is learning. This is especially true for developing jobs that are as yet unclear, such as excavating and masonry. It is also true for clear jobs, since they too can change in the course of construction as conditions, new judgment and redesigning emerge.

Thus, another discomforting consequence of being instructed and yielding control for the period of traineeship is the opening that it gives the sub in redesigning. The patsy's recourse to power on the taste level is still strong, but technically his loss of control can be quite disconcerting. He must either counter for reasons of taste, or suspend the apprenticeship and exert his generalling power "no matter who is right." The latter seldom happens; the former is often the control strategy. (For example, when the rocking crew complained about wrapping the windows with metal strip and instructed the patsy how windows are usually trimmed, or when they instructed on the benefit of stipple walls as opposed to smooth, the patsy's only leverage was to say that wrapped windows and smooth walls, while technically not the best, were what he wanted as they looked best to him.) When the crew is both technically and tastefully right, the patsy's strategy, after he complains, is to apologize and retreat to the student role. "Can't I just go on learning," so that the crew is not offended. (An illustration: when watching the mason float the fireplace slab as he was explaining about it, the slab looked too dry to me, so I said, "God, is that cantilevered slap going to hold—it looks bone dry?" "What do you think we are, stupid?" Then he explained how they draw the water out with a flat board.)

Sometimes the profit of instruction is having the sub come up with a technical redesigning which allows the patsy to implement an impossible taste consideration. (Thus, when the heating contractor, explaining the theory of cold air returns, discovered how to relocate the planned one, the author then was capable of redesigning the "used brick" wall to suit his taste. This occurred several times with the foundation contractor who said, "I'll talk anytime to anyone about building a house" he so loved to teach. He was a constant teacher and consultant.)

A final consequence worth mentioning is that crew and subs alike, instructing on their own volition, usually take pride and joy from it. This gives the patsy another ground to retake control of the job, as a good student who has learned and applied his lesson. The sub enjoys this product of his efforts; the patsy now practices what the sub has preached.

Besides the student role and its beneficial derivatives, other sources help the patsy regain generalling control. As discussion of the information-gaining process in Chapter III showed, the patsy learns how to inspect, evaluate, and administer jobs from being instructed in their techniques and details. These abilities allow him to take cogent control, through "trained" judgment as he tries (and confronts) to make ideas fit actuality. The patsy no longer acts with doubt and fear of blunder when questioning about details of the ongoing job. These abilities clearly give him the edge in power symmetry. Often this regaining of control comes back in the nick of time to curb or redirect the crews' judgments and redesigning.

Incompatibilities: Compatibilities and incompatibilities on the job affect morale; in turn these can affect quality. Since one criterion in choosing a sub was the potential compatibility of working with him, now the patsy has his chance to actuate the choice. The main source of compatibility is how the patsy supports the crew or sub. He may do it verbally, as by encouraging, praising, consoling for difficulties, resolving differential judgments, and playing student. He may also do it actively by pitching in and helping, giving and receiving gifts, generally working together and jointly handling errors.

We have just seen how differential judgments are resolved and result in a degree of compatibility that can easily help or hinder the quality. The patsy may take control too abruptly, or take it back too quickly from a lesson, and hurt the crew's or the sub's feelings. (The taping crew, annoyed at having a suggestion regarding painting ignored, slopped mud around and got it on the redwood door trim. "They were angry, out of shape, and made a mess of things.") Mostly deferring and joint resolution results in ensuing compatibilities, which foster efforts at doing the best possible work toward the mutual agreement, (In working out the best shade of mahogany stain for the hardwood floor together with the sub, I encouraged an excellent job that both of us were delighted with.) Verifying crew questions on details, to bring out both the crew's as well as the patsy's opinion, has the same effect. The crew feels supported and respected.

Praising is highly motivating for crews and subs when their job is highly skilled. This applies mainly to finish jobs, such as tile work and masonry. Flattery, by asking the crew to work on other jobs, is particularly effective. Rough-in jobs are harder to praise since the focused aspect is on technical skill, which is usually not that difficult. (To tell the roofer his work looked great did not help much with his motivation, his job was so routine. In contrast, praising the perfection of the taping crew regarding the smoothness of the walls was quite gratifying to them. They impressively put themselves out beyond the call of the bid.)

As a general, the patsy can always help quality by never missing a chance to console and understand for self-blamed and corrected mistakes, to humor in frustrating circumstances, and generally to encourage. These efforts smooth the rough spots in work, and help keep the crew's tolerance and make them strive to do quality work in a tight situation. Many is the time they are backed into awkward bodily positions (as with plumbing and wiring), where they literally have to fight through a detail. At other times they are working in a lonely spot on the premises, and during the intermittent watching, the patsy does well to chat and encourage so that the crew does not feel out-of-it.

Working actively with the sub or crew is stimulating to their good work, so is working compatibly side by side on different jobs. The patsy's always being ready to stop what he is doing, to help the crew for a moment, as handing a tool or moving material, again motivates for the crew to respond with quality work. Pitching in with these helpings takes the incompatible edge off of close watching, which can undermine the patsy's power.

When compatibility starts to peak, it is not unusual for the patsy and crew to reciprocate with gifts. The patsy allows some minor lifting of materials or tools (the unofficial reward system). Another form of gift is the lining up of another job for the sub. The crew may give free extras, behind their bosses back, although they are minor to be sure. (For example, the electrical crew threw in an extra circuit breaker, the plumbing crew an extra cleanout, and nothing was said about either on the bill.) Crews have a varied leeway to do this during a job, when motivated to express compatibility with the patsy, viewed both as general and as a person. Usually the extra gift comes from their own redesigning based on empirical know-how from fixing other houses and construction. They consist of little, preventative trouble-shooting items, which will save the patsy trouble and money in years to come. To be sure, they also increase quality.

Handling errors is another activity that the patsy must deftly general, so as not to impair compatibility by lowering motivation. Thus, pacing a correction is vital. The patsy waits for the right moment to suggest a corrective change, usually when the crew's work is at a convenient breaking point. The patsy finds it is better to confront crews directly than to go over their heads to the sub, since that adds an unnecessary put-down. The crew has to fix the error anyway, no matter who knows.

Inept crews either must be stopped cold or left alone to try and work out the problem, but under close watching. For the patsy to butt in, in the middle, is even more confusing to them and affects quality. Often crews chastise each other for errors, so the patsy has the advantage of agreeing with one that the correction should be made, while modulating or mollifying the attack of the chastising side. The patsy can have it both ways.

Some crews are adept at listening to the patsy's demands for corrections that are not based realistically on the gradual outcome of the job, and then going on as they know they should. Thus they do not allow the patsy to blunder away the compatibility of the job. Crews are accustomed to such unfounded requests from owners of custom built homes who come to the projects to watch how work is progressing. The patsy is careful not to fall into this category and thereby undermine his generalling powers.

Starting the job and then running it are the first two phases of generalling the quality of a job. Bringing it to a satisfactory closure is the last phase. Aspects of closure are discussed in the next chapter. Another aspect of closure, only touched on in the next chapter but already dealt with in detail in Chapter V, is the temporal closure of a job—assuring that it is done on time.

VIII

Closure: Temporal, Quality, and Financial

Generalling the closure of a job is a complex activity, composed of articulating the many factors that press for closure. As indicated in previous chapters, each job generates its own pressure for closure. As it approaches its finish, it needs finally to be finished—to be over with. *Jobs are self-limiting:* they can only go so far and then they are finished. If they are a quick or base job, of course, there is pressure for them to be completed so that a subsequent job can begin. The patsy as general must articulate these job-closure pressures in terms of his overview of the project, keeping the project moving, and timing quality and paying for each job. Additionally, he is under pressure for closure from the sub and his crew who, as the job nears its finish, are more drawn elsewhere to new jobs and pushed to making the contract pay regarding their time and costs.

In this chapter, I shall discuss how the patsy guides jobs to closure in line with these pressures. First, I shall consider the pressures to close because of job articulations and subs being drawn elsewhere and "making a contract." I then discuss preparing for the inspection, negotiation, leave-taking and comebacks necessary for finishing odds and ends, and for the tension released upon closure. Last, I discuss how the bill is settled. These sections together deal with a complex, cumulative process. As pressures mount, civic inspects are hurriedly prepared for and the patsy himself inspects; meanwhile the crews and sub are constantly taking leave, while the patsy prevents the leave or negotiates for "comebacks" to finish up, and he uses his power of settling the bill to keep his edge in the power symmetry for securing an acceptable closure. His fail-safe measure is to demand corrections through legal statutes on subcontracting guarantees.

Pressures toward Closure

The three strongest pressures toward closure are the articulation of successive jobs impinging on the patsy, making the contract within specified time and price, and elsewhereism impinging on the sub.

Elsewhereism is a far stronger pressure on the sub than *making the bid.* His making the bid is a composite of achieving the desired equality in the requisite time for the price, then going elsewhere to the next job. Either time or quality can cost too much, but usually it is time that runs out, as time is harder to pre-calculate. Overruns on the time mean excessive labor costs and loss of time on the next job. To offset this danger, the sub will usually figure into his bid a temporal cushion and charge for it unbeknownst to the patsy. With this cushion resolving his temporal problems, elsewhereism is given a clear path for pressuring the sub toward closure.

The author continually rediscovered this cushion in many jobs done by subs. The foundations were done in four, not five, days; the tile sub put in an additional tile floor within the time he had planned for the original job; and the heating crew "finished faster than they wanted." Coming toward closure, the sub or crew is continually warned about not finishing on time. At closure, I discovered in virtually 90% of the jobs that I had paid for time not needed. Thus, the sub's time is used on jobs elsewhere. The patsy has little to challenge—the sub is making a bid, not working on time and material. In one case, when confronted, the sub retorted, "We usually don't get work done fast enough to even make our wages, but we made them here." This disavowal strategy was hard to believe but closed the issue immediately. Both knew the sub was making his money too fast, but he had the edge because of the contract bid!

The few contractors who failed to make the bid temporally irrespective of or because of no cushion, resolved the job on a comeback basis. They took their crew elsewhere in the face of protests by the patsy and sent crews back at odd or slack hours to finish up. They were no longer devoted to the patsy's job on a full-time, but rather on a spare-time basis. They generated many delays. (I shall discuss leaving and comebacks further in the sections below.)

Elsewhereism increases in strength as jobs near closure in order for the sub to compound his profits and meet his other demands. In "beating" his bid, the sub attempts to compound his profit by moving the crew to the next job; for a few hours or days their work is virtually paid for twice. However, meeting the demands of promised time and need elsewhere is stronger anyhow. The promise to be elsewhere with crew and/or equipment is the prevalent condition pulling the sub off the job. "I hope we get this form finished and inspected by Thursday afternoon and qualified because we have to be on the next job Monday, and they have been waiting a week!" This force virtually overrides any edge in power symmetry that the patsy may have to keep the sub finishing, if his finishing is a reduced work load such as trimming, grooming or cleaning. The patsy's withholding payment for inspections is not a strong deterrent since the sub will always come back for odds and ends in order to have his work passed and get paid.

If finishing is still a full-time job, the sub may use the strategies of intermittently leaving the job or forcing poorer quality by rush work. The former strategy refers to leaving the patsy's work to start the elsewhere job, then returning full time to finish, thus hooking the other client while only delaying the patsy a day or so. The latter strategy refers to asking the patsy to settle for a less than perfect, but still quite suitable job, (The foundation contractor, to save a day, wanted to pour concrete on two successive days which meant a cold joint in the concrete. This was O.k. but not as good as if there were no joints at all.) Fast finishing can be a panicky, frantic situation, diluting the patsy's ability to make judgments, while everyone is turning in all directions. City Hall inspections, when they apply, are a good excuse to use for the patsy's slowing down the quick finish. Otherwise, he again must resort to comebacks and guarantees when forced to accept less than adequate work in the heat of a closure rush.

But not all subs are pressured toward closure to the point of restraint. Their usual pattern is slowly to cut down on work, causing small delays because of their inability to get crews from new jobs to come over and finish up. Thus, the patsy's problem is to pressure the subs toward closure because the next job must start, so he baits them with quick settlement payments. This is very effective, since most subs can always use liquidity for immediate expenses, because most of the sub's accounts drag out settlement because of their characteristic thin liquidity.

In sum, these pressures upon the sub and the patsy to close jobs then force inspections, premature leave-takings, comebacks and the use of settling bills to slow or quicken closure. Hence, the next discussion is how the subs are generalled by the patsy to properly achieve closure. Reciprocal forcing is clearly the generalling problem to be handled with closing the self- limiting job.

City Inspections

The patsy's inspection of a job is, as we have seen, a continual aspect of generalling from start to finish. The city inspection required of several jobs—plumbing, heating, wiring, masonry, roadwork and so forth—is a crucial part of closure. Not to pass means the articulation of the project cannot continue and the sub cannot be paid, and if he has left then he must return from elsewhere to correct aspects of the job that are not up to code. Thus, passing inspection is of deep concern to both patsy and sub. So they work together to succeed at each stage of the inspection process—preparing for, calling for, and passing or not passing the inspection.

The first two stages are ones of delicate timing and approach, the last stage is one of waiting for the outcome. Since the patsy and sub have little effect at the moment of inspection, they pour their efforts into strategic preparation and timing the call for inspection to effect a favorable outcome. They usually succeed, but can easily fail (through forgetting a detail) on minor offenses to the code, which require corrections and reinspections until a job is passed. There is never complete failure in single family residences. (For example, on a cement block wall, the mason forgot to fill one cell, which he had to come back and do before the wall was passed so it could support a brick wall.)

Preparing: The crucial problem to keep in mind is, "Will the inspector pass the job?" The typical subs' discussions predict the passing on two grounds: whether it is up to code and whether the inspector's personal life is O.k. that day (whether he had a good dinner the night before, slept with his wife, got up out of the right side of bed, or just feels good.) Personal life is at issue because "If he is just plain angry and wants to take it out on someone, he can always find something wrong." Such caprice flows from personal conditions, when jobs up to codes are not passed, in the view of sub's experience. To cope with the personal circumstances, subs and crews will dope out at length the personal requirement of inspectors as they have experienced them in other jobs and heard of their quirks from friends. They build these requirements into their preparations. For example, the *clear goof* is a strategy for the inspector who is a frustrated builder, or architect. He has to have something to correct so he can apply his expertise; if he finds nothing he will find something unjustifiable which can be trouble. The goof also deflects his attention from something that really is not up to code. Giving him one juicy correction which can easily be fixed uses up his "inspection account," makes him happy so he can pass the job feeling he has done something. Thus the sub and patsy control both the error and the inspector. These kinds of goofs are easy to plant, like putting a foundation bolt in the mudsill upside down.

When a job is not quite up to code, often it may be because stringent code requirements can easily interfere with details of taste—yet no structural damage is at stake. For example, lowering a ceiling in a breakfast nook to make it cozy is not up to the code requirement that all ceilings must be eight feet high. In cases like this the sub and patsy are virtually forced to get a job passed that is not up to code. Three kinds of strategies used for this purpose are omissions, commissions, and fudging—the latter being the standard one. The patsy learns these strategies through listening to tales of complaint about other inspections, as well as by working with the sub or crew to get a job passed in spite of a few flaws.

Omitting is a tempting strategy. After all, the inspector can easily overlook small details as he scans or becomes absorbed in other details. Not switching wires properly in a room, not overlapping the corner of double plates, forgetting a bathtub cleanout under the house; all can easily be omitted and overlooked. But who takes the responsibility if caught? If the crew is caught, they are seen as bungles by their sub and the inspector for goofing on such a small item. It could affect a crew-man's apprentice status. If the crew shares the responsibility with the patsy, which they try to do, they are all seen as poor workers. If the sub joins in, then the inspector sees collusion to put something over on him and is hard on his inspection. He finds several nuisance items with which to goad the patsy, taking up time and energy. He also becomes wary of the sub on all future jobs, which is annoying for the sub when he tries to make contracts pay. Thus, the omitting strategies, although easy and apparently inconsequential, are usually not used because of the consequences and unwillingness of any one person to take responsibility for oversights. "No, you better fix it," is the consensus. If anyone takes responsibility on his own (crew, patsy or sub) it can also be costly, since he would have to pay costs for the overtime to fix the omission. The patsy loses his edge in power symmetry in sharing this responsibility since, by contract, the sub must get his job past the inspector anyway. The patsy is only pressed to give in by the timing of other jobs that require closure of the current one, say, the basc job of wiring so walls can be closed in.

Committing errors as a strategy is a variation on the goof. The latter variation is to give the inspector a bone to chew on, whereas the former is to deflect his attention from a real problem. For example, when a structural feature for taste of the building forces a clear violation of the code—such as lowering door heights two inches or having beam edges below the eight-foot ceiling height—it may be best to soak up the inspector's time and energy for criticism on an item that *can* be corrected: such as the height of risers in a Staircase still only roughed in. Inspectors can only find so much wrong before they are worn out, must go off to another inspection, or appear to be unjustifiably abusive. This strategy is only infrequently useful. It is perhaps used to cover the one or two real code deviances of a building. There are, of course, usually a few where the code is stretched for keeping taste consistent in the building—such as making a firebox an inch too small so it will fit into the space in a wall.

Fudging strategies are the stock in trade of all crews and subs. There are typical points in all jobs where the code may need to be stretched a little according to common sense. The basis of a fudging strategy is to make a slightly illegal detail look legal. For example, if the foundation is eight inches over the setback (not much in suburban construction), the surveyor may suggest moving his boundary stake eight inches when locating the foundation on the lot. It would be a nuisance and a hardship to spend time to go through planning for an eight-inch variance next to a three-foot Pacific Gas and Electric Company easement. Another instance: the plumber makes the line into the septic tank short so it appears like a T, which it is not. The patsy usually joins in with taking responsibility for these fudgings since they are in the service of keeping the job moving and tasteful, when they could be needlessly halted at considerable time and expense.

Sometimes code requirements conflict, and fudging requires deviating from one requirement to suit another. (For example, in order to make the hearth a required 20 inches across the front, it was necessary to cantilever it five inches over the cement bearing wall, a change which was almost impossible to spot.) The complexity of jobs often requires fudging somewhere along the line to make it all fit. Inspectors are aware of this too and can overlook with ease nonobvious transgressions when their personal lives are going all right.

Calling for Inspection: Preparing for inspection is worked out together between patsy and crew or sub. This can also lead to designating who calls for inspection, but not necessarily. Both patsy and sub have different pressures on them for closure, which affects their timing of the call. The sub is in a rush for money and is going elsewhere. The patsy is articulating several jobs and most crucially does not want an unfinished one seen by an inspector. The unfinished job is likely to be unprepared—particularly unfudged. Inspectors delight in making comments about how a job in progress must be done to pass a formal inspection. They can spot details they would not see at inspection time, *nor* before inspection would be allowed on the premises to see. It is up to the patsy, therefore, to keep his edge of power symmetry on who will call for the inspection and when will he or the sub invariably call as fast as possible. (The heating contractor did call unawares. The inspector found the heating fine, but complained of many framing details not prepared yet for his eyes, still early enough to call a halt to—much to the author's chagrin.) It is difficult for the patsy to be told—as a patsy—to put in kicker blocks for wall bracing when he would do them in due course anyway. It is worse for him to be told to raise a door header only half-framed when, if it had been framed, it would be too much to do. The unprepared for, uninvited inspection denies the patsy the *fait accompli* aspect of committing and fudging strategies, where corrections are a hardship relative to the minor degree of deviance.

Passing Inspection: As I have said, all jobs pass inspection; it is just that some pass immediately and others do not until the inspector thinks they are done up to code. Those that pass on the first day are either O.k. in the inspector's view or he lacks the confidence to stop a job because of borderline violations. Not to pass a job and thereby delay construction is costly and frustrating, which puts pressure on an inspector. So he must be quite clear, and shies away from debatable transgressions of the code when he is angry. He points out the problem and nothing more. The patsy acknowledges the possible transgression and nothing more. (Frequently the beams of the ceiling were seven feet two inches when required code is eight feet, but the surface was so small compared to the roof deck that the inspector only pointed out this apparent transgression.)

When a job does not pass on the first inspection, the inspector has the power to *force* changes in construction. The degree of his forcing varies from the extreme of resubmitting plans and rebuilding an aspect of the house (e.g., different window structure and sizes) to requiring unnecessary small changes (e.g., raising front door height two inches). Forcing toward the lower, extreme end tends to be necessary and the leverage that the inspector uses is a technical rendition of the code (e.g., each room requiring a certain amount of window area). Forcing toward the small-change end can require technical change (e.g., more tabs are needed in the masonry wall); it can require taste changes of a redesigning nature, leveraged by a technical requirement (e.g., raising door heights). The latter type—forcing taste with technical power—is often based on imperfect or fictional renditions of the building code. This outrage sometimes can be handled by the patsy through a reinterpretation of the code for the inspector who is so carried by his own taste needs.

The more typical corrections fall in the middle of the continuum and the forcing generates negotiations as how best to correct violations acceptable to the patsy, inspector and sub, so that no loss in time or money is incurred. Usually the sub will handle the negotiation, because of his experience with his type of job, unless he becomes so defensive *or* carried away elsewhere that he leaves the patsy to work out a new closure. The favorite strategy of subcontractors is *cycle suggestion.* The sub carefully makes simple, tacit suggestions in the course of discussion and as the inspector gets the idea, *he feels it is his* and suggests the solution. The sub and patsy need only agree to it which makes the inspector feel he has done his job. This strategy is preferable to trying to make him accept a solution from the sub, which, in view of his power, may easily make him defensive and require excessive additional work by the sub and patsy. Making the suggestion also makes the inspector very compatible if he is a frustrated architect or builder who needs to impose his thought on jobs. Cycle suggestion works well with simple ideas and mediocre people.

The patsy, if he is to do the negotiation, may also use cycle suggestion or use it under the tutelage of the sub. But he can also use *blaming*—he has been a victim of the sub's failure—as a strategy to secure for closure a minimal time and cost correction. Some inspectors feel sympathy for the poor patsy in the hands of a fess-than-scrupulous expert. (Regarding a sewer hookup, when the inspector heard what sub had violated the code minimally, he said, "Oh well, forget it." This indicated that an arrangement had been made between the inspector and sewer contractor to cover many such minor violations, though the patsy will never know. But it is clear the inspector did not want this sub to undo his work for a minor correction.)

When the inspector demands full compliance with the code, negotiation reverts back to the patsy and sub approach to the correction. And they determine to what degree the compliance will be fudged, and who is responsible for it. As said earlier, the patsy will be under pressure for closure because of articulation and the sub in order to keep costs down and to meet demands from elsewhere. (In the case of placing more ties in the masonry wall, the negotiation revolved around to what degree they would be sunk in the wall, and to what degree would they do a realistic job. The deeper the tie, the more time and money; yet the inspector would never know unless he pulled out each tie. So from the overview of general, the patsy settled for a less-than-deep addition of ties to keep the job moving (cut down on the delay) since the requirement was excessive anyhow, also to keep the complaining sub quiet about his marred record. In this case it was the first inspection he had failed in ten years.

In the end, the jobs that do not pass on the first inspection are often a result of premature leave-takings by the sub, so let us turn to how the patsy generals this conflict.

Leaving the Job

Because of the pressures toward closure, there easily arise differential judgments by the patsy and sub of when closure occurs—thus when the sub can leave the job, whether for good or for comebacks to take care of odds and ends. Besides the pressures of articulation, stray elsewhereism demands, and costs—which tend to generate less than adequate quality—sometimes the patsy must be sure that the sub does not overdo a job, thereby reducing its quality. This difference of judgment—especially regarding quality—generates a temporal pacing problem that requires the patsy to general to his utmost, using his powers of inspection and payment. The patsy must make closure judgments that he can stick by in the face of counter definitions by the sub (or the crew). When closure is not agreeable to both sides, the patsy will usually be faced with preventing premature closure or delayed closure by waiting for comebacks. (Overdoing the job is infrequent, but can harm the quality of finish work such as painting or tiling.)

The general rule is to never let a sub or crew go until the job is correct by the patsy's standards and possibly by city inspection, even if they have to stay extra time. With a contract by force, it is their time not the patsy's, unless extra work is involved. The patsy can apply this rule to those subs who state that they are finished and wish to be checked out on quality and closure. If they are not finished, the patsy demands that they stay and finish, even if this means overtime. At minimum, if he cannot hold out against elsewhereism demands, the patsy negotiates comebacks with the sub to articulate closure with successive jobs. When the sub simply takes over, the patsy starts with negotiating articulated comebacks to what he defines as an unfinished job. Quality and time clearly intersect at the sub's leave-taking. And who has the edge in power symmetry is decisive in favor of either the patsy or sub regarding when the sub takes his leave. The patsy has more power in demanding corrections for a city inspector than for his own judgment.

If the patsy's judgment is "no closure yet," he can use any or a combination of several denial strategies to try to prevent a sub's departure, as based on his powers of inspection and payment. His most benign strategy is to emit *no sign of closure* (sign denial); his most extreme strategy is to demand a redress of a *completely wrong job.* (For example, the roofer tarred in the skylight wrong and I demanded he rip it out and do it according to my imagery. The crew could not leave until the job was "right" and feel safe on closure.) Thus, these two strategies generate an inability for the sub to feel closure from the patsy and thus for himself. There is tension—not the tension release that comes from good closure.

Two other denial strategies by the patsy also have this effect of forcing the sub to feel no closure. One is emphasizing, often constantly, to the crew or sub that the job still has a ways to go, just as they appear to be finishing. The other is extemporaneously to help prolong the job, in any fashion, so the crew will stay and finish. (Once, I quickly built a plywood roof over a mason caught in the rain so that he could finish up at a sound pace with no comebacks.)

The sub's most frequent counter-strategy used to offset the patsy's attempts to prevent—in his judgment—premature leaving is to promise the comeback, and if necessary to schedule it and explain how it will handle remaining work on an articulated basis. (I deal more fully with comebacks in the next section, since comebacks versus preventives to leave become a tug-of-war needing fuller attention—generalling, power symmetry and elsewhereism being the main ingredients). Other counter-strategies are cutting the patsy short, pleading elsewhereism, disappearing, teaching the patsy to finish himself (if he will), and the crew referring the patsy to the sub. Cutting the patsy short is an easy technique for crews and subs who are not great talkers anyway. They either do not respond or respond cryptically to the point where a communication gap develops on closing imagery; and their leaving is based on their judgment. The patsy can only demand comebacks.

Pleading elsewhereism we are familiar with: "Bye-bye, I have to go to another job. I'll see you in a couple of weeks. I'll let Herb come back up and finish." Pleading elsewhereism is covered by the inevitable promise to come back. Sometimes a sub will try to teach the patsy how to put finishing touches on a job, such as wiring in one light fixture. This seldom works because of the shift in responsibility and liability if the patsy accepts. It works only under great pressure from articulation problems. Crews will try to refer the patsy to the boss when, in fact, they themselves should verify closure with the patsy. In any case, with or without these strategies, the one of last resort is to disappear during or at the end of the day, only to come back on demand. The ease of disappearing by not returning renders the sub the edge in power symmetry at this juncture. So let us turn to the comeback phenomenon.

Comebacks

The main sources of the tug-of-war between comebacks and preventing leaves is, as we have seen, the generalling of articulation, costs and demands elsewhere. The edge of power symmetry resolving this conflict is based on these pressures and the ability to use the strategies to enforce each side—discussed just above. But also, patsy and sub can agree when necessary to a comeback basis for finishing a job. Comebacks—up to years—are also implicit in every contract based on the guarantee, even when closure is achieved before leaving. Here I wish to discuss the nature of comebacks for closure *per se,* whenever they are used.

Comebacks are used to gain closure in two ways: fixing a job and tapering one off. Jobs are tapered off for several reasons that the patsy can easily agree to, and finds hard to prevent even if he desires. A typical closure pattern for crews and subs, whether they are harried or not, is to leave one *unnoticeable thing* undone. When it is discovered days or weeks later, the oversight requires a comeback. Even if payment has been made, the comeback is part of the guarantee. Running short of materials also forces a comeback whether the patsy likes it or not. This is caused either by the sub ordering too closely: he omits a board of sheetrock, or a special item, such as three corner tiles. Again, the patsy can only yield to a comeback closure.

The nature of the job and of the day's work also may force comebacks on the patsy. Getting done *by the end of the day,* after which a crew is scheduled to go elsewhere, can result in finishing a few odds and ends "in and around" the next job when they are able to come back.

Certainly the few details do not merit bringing a whole crew back. If articulation presses, the patsy can offer overtime pay or plead with the crew to take an extra hour to finish up, which they may if they need to feel closure and start the morrow afresh. If this fails, the patsy when appropriate may use a listing strategy for complex jobs. He goes over the job carefully to list what remains to be done, so that everything will get done without too many callbacks. Also if comebacks are performed by different crews, or often the sub himself will do these odd but unfamiliar jobs, there will be an understanding of what needs to be done. It can be done as fast as possible since several little details can be planned for. (This worked very well with plumbing and electrical finish jobs.)

When comebacks would be for too simple a job—wiring in one more light fixture, or dropping the trap in the firebox floor of the chimney—the crew may prevail on the patsy to finish it himself rather than demand a comeback. They may even wish to train him for five minutes. The patsy must judge at this point whether to take responsibility and release the crew without any planned comebacks. He may agree to the finishing himself, when he sees the crew pushed by their sub to finish and move elsewhere.

Comebacks for tape ring-off jobs win trust when met on time; comebacks for fixing mistakes do not, or only barely so. These latter only reduce the tension generated by a low-quality detail. Thus, subs usually "cheerfully and willingly" promise these comebacks—especially those occasioned by the building inspector—immediately as a reduction strategy. Such response "steals the patsy's smoke," until (and unless), of course, the sub *does not* come immediately. Perhaps the deciding criteria for actually meeting this demand immediately is awareness of the patsy's articulation problems, and whether the bill has been settled. (A sub corrected a door within twenty-four hours because it held up progress, and he was not paid. The tile contractor responded to a loose tile immediately; the cabinetmaker to a single oven cabinet which should have been double. The cabinetmaker needed payment immediately as promised, because he had been paid and construction was not hindered,)

Coexistent with immediacy in fixing is a *faulting* strategy used by subs. Just as fast as they respond to fixing for closure, the subs try to afix blame for the mistake and perhaps pick up an extra charge. They quickly review their contracts and the ongoing situation to see if the patsy was at fault, or shared the fault. Did the patsy walk on the tile too soon and jar one loose? (The cabinetmaker said the author had ordered a single oven cabinet but, to the latter's denial, the sub ended up sharing the cost of immediate rebuilding of the cabinet.) Seldom, however, is the patsy at fault; the fault typically is by an oversight, negligent crew, or is a normal error. Tiles come loose from time to time. The crew did forget to fill the concrete blocks every four feet, or put enough wall ties into a brick wall. Faulting gets the sub little or nowhere in the edge of power symmetry. The job must be set straight, in any case, per guarantee, and bill settlement made so that the project can proceed. Fortunately most mistakes are simply and easily corrected, and most bills have not been paid; so unless faulting is securely against the sub, it only creates bad feelings toward the patsy which undermine the two main consequences of closure: a happy tension release and the pacing of settling the bill. Thus subs usually raise no question. Indeed, they often err in the other direction before mistakes arise by saying, "Call us no matter what kind of problem you have and we'll fix it." This promising of a comeback "in case" gains trust.

Tension Release

Generalling of the leaving and the comebacks for jobs near closure builds up the already existing tension; closure releases the tension. The ups and downs of gene railing jobs exist for all jobs in the course of construction. For dangerous jobs, base jobs and finish jobs, they can be quite deep and steep. One minute the patsy is depressed at timing and/or quality, the next elated—depending on how the job is moving and how he is running it. Closure brings an end to these modd swings, with a release of tension about the job forever. This release is manifest in three ways: self-other, job focus, and forgetting. The first tow may or may not occur together, the other always follows.

Self-other manifestations are diverse ways of indicating release and relief. The patsy and crew or sub may express job satisfaction and/or excitement and delight toward themselves and each other for their achievement. They make jokes and tease each other about the difficulties encountered in reaching closure. They anticipate payment and paying for a job well done. They chat amicably about the job and what job is next—elsewhere. The patsy becomes concerned if no job is elsewhere. They congratulate each other. They mention each other's personal lives.

These expressions easily move into focus on the job itself. "It's beautiful," is typical—"I'm satisfied," "Good work," are also used by the patsy.

The patsy in particular responds to the transformation that closure brings with respect to what the project was before the job started—for example, a bathroom or kitchen without tile compared to its being tiled completely is a striking, productive change. He is likely to praise the job, hence the skill of the crew, especially for finish jobs. He expresses appreciation, and the crew accepts and returns it. They talk about the patsy getting good work like this in the future and referring their work to other people. With respect to finish jobs, the patsy and the sub talk about how houses are sold because of beautiful jobs.

Then . . . the sub leaves, the job is over, the patsy devotes his attention to articulating the next jobs, the sub to a job elsewhere, and what was a striking transformation a few hours before becomes *an assumed fact about the total project—and is forgotten* as a specific job almost as intensely as it was focused on before closure. Unless, of course, a guarantee problem arises. For the patsy, one more joyful step toward a complete house has been taken. The closer toward completion each job brings him, the greater his joy.

Settling Payment

The last step in closure is settling payment with the sub. This may come as part of the tension release, but usually comes at a later date, because the normal pace of settlement is thirty to ninety days. There are three main issues involving settlement: its pace, its mode, and "gouging."

Pace: The pace of settlement is usually normal in the sense that the patsy receives a bill soon after closure on the first of the month and has thirty or sometimes ninety days and with some subs, need never worry. Occasionally payments are needed *immediately.*

Normal pacing is built around the patsy's ability to make payments. As we saw in discussing bidding, subs' concern about payments made them query about how the patsy would pay, unless the patsy told them. My project was backed by a building loan which was issued in five payments linked with the course of construction, which in turn determined normal pacing of settlement for the subs. The author's job as general was to keep the project moving toward each payment point—e.g., after foundations, then subfloor, then roof, etc.—so he would receive a loan payment to meet normal expectations of the subs who would have submitted bills to that point. Obviously I had to keep articulating so these payment points were reached before thirty-day settlement dates expired. If I missed this pacing, then explanations and requests for waiting were due some subs; if I did make payment points in construction on time, nothing was said, bills were settled on time by mail. This patsy had a good record on articulating construction with loan payments and bill settlements. Many subs were surprised to receive money on time instead of "having to wait months," as they do with many general contractors. One sub said out loud, "We better finish this job, Joe, there's money here." Thus, this aspect of the patsy's gene railing *reputation* had traveled, and aided obtaining closure in many jobs in the face of strong elsewhereism. In the final analysis, subs work for money, and such a reputation travels fast in the neighborhood bars after 4:30 PM.

Had I indicated the slightest doubt about payment, pacing job closure would have been difficult in many cases, since generals elsewhere who give them many jobs—even with delayed payments—would outweigh my demands. Slow payments are risky for a patsy, especially over sixty days. But houses move fast, so a patsy would have to be poor at generalling to foul up the structural linkage between course of construction, job closures, loan payments and bill settlements. I always kept subs advised of when the next loan payment was due, to allay concerns. Delayed payments are more a consequence of poor bookkeeping and disbursal, and overextension of liabilities.

Sometimes I sent a check so early that the sub was not set up to receive it and I had to verify receipt of uncashed checks. (For example, the electrical contractor deposits checks once a month, so let a check sit three weeks which, fortunately, he did not lose. One roofer was so accepting of the building loan source of money that he never filed liens. If he missed a disbursal out of one loan payment he waited until the next, or waited until the home sold. He did not care about the patsy's early payment.)

In short, normal pacing of settlement requires little or no relating between patsy and sub, except perfunctory advice of loan payments and settlement dates. *Immediate* payment needs break through normal pacing and may generate considerable patsy-sub interaction. The patsy may be the author of immediate payment for several reasons. He may offer it to motivate immediate closure. He may use it to get rid of an obnoxious sub. (For example;, I paid the tree man off quickly before he made more mistakes.) He may wish to settle quickly before a sub may add a legitimate extra to a contract price, hoping he will be delighted to get paid and forget the possibility. When noticed by the sub, the quick payment gives rise to discussions of delight and thanks.

The two sources for subs requesting immediate payment are fear and need. (Only the floor contractor had any fear that he might be left out.

He was the last man on the job, and it has been his experience to have the last loan payment run out just before he was paid. He certainly did not wish to take chances waiting thirty days, so pressed for his money in partial payments even before he finished.)

Weekly liquidity need is a frequent plague among small subs. The crew has to be paid on Friday, so the sub hustles where he has the most likely chance of obtaining immediate settlement: the patsy. Often subs keep their crew on a week-to-week basis, and when they fail on Friday some crew members may go to another sub.

Subs have several strategies to hustle partial payments before closure or quick payment after. They bargain for partial payments at each stage of the job, such as plumber, electricians and the floor man. If they have put *it in the bid,* they bill for it. If not, and they need the money, they make a *special trip* to ask for a partial payment and explain their problem. "From now on I am not going to do any new construction until I get three-fourths of my payment." In explaining, one sub said a back-up strategy is to *walk off* the job before closure to force demand. Seldom do subs walk out when a building loan is backing the project, yet partial payment is possible to them when a loan payment has been made.

Another strategy is the *spot phone call,* catching the patsy off guard. ("Hi, we are having a campaign around here to get money—how about your rough-in payment?" This call was from the plumber who knows loan payment schedules very well.) Facetious excuses, particularly travel, are also used. ("I need a little money out of you because I have to pay Bob. He went to Reno and he's got to recoup his losses.") The quick payment is for the sake of a crew member, not the sub. (In this case the crew member had done a fine job, and the patsy was glad to oblige, not really knowing if the money did reach Bob.) The sub may even *discount* the bill to obtain early payment, which is excellent bait for the patsy. As we have seen, a sub may *call for an early city* inspection as a way of getting immediate payment for closure. But these untimely inspections are likely to arouse the patsy's ire and make him wish to delay the payment. In hustling quick payments, the sub never explains his need as necessary for gaining profit; it is always for liquidity to retire financial demands elsewhere: "I'm short of cash." The sub usually has many uncollected bills and needs money no matter what it represents regarding profit or return of capital. Under the onslaught of these strategies, *the prudent patsy must remember he is paying for a job, not baiting out a sub.* He must be careful when a partial payment means delay to closure. After closure he must be careful how he is distributing money, so that the quick payment will not disrupt his own needs for liquidity.

Immediate payments may not be trusted to the mails like normally paced settlements. Once the patsy agrees (which he does if he can), the untrusting sub makes a date for a trip to the building site within an hour to the next morning. The sub never lets the situation cool longer. "Gordon, shall I put the money in the mail or do you want to pick it up in the morning?" "No, no, no, I will pick it up in the morning, don't put it in the mail." Surety is money in hand.

Mode: The mode of settling the bill follows from the pace. If the pace is normal, then bills are settled by mail. Subs will often query the patsy on where to send the bill since, of course, the new home is not yet an address. If the bill is satisfactory, the patsy pays by mail and nothing more need be said. If not, he queries. Not infrequent strategies of subs are to slip in extras or forget credits. When querying, the patsy must be prepared to *argue* his questions. Another counter strategy of the patsy is to return the lump sum bill for itemization so that he can compare it exactly to the bid and see if an extra or credit was warranted. Small bills can creep in as, say, extras for a permit for a special item or time on the patsy's behalf that were not included in the bill. Similarly credits can be owed, say, for a change in style of a lavie that was cheaper or "My God, you charged me for all that firebrick and it's still sitting outside, it was never used." Claims of excessive time are not the patsy's responsibility.

The sub hopes that the patsy will simply pay, but under the patsy's query and argument he must reduce the bill with apologies if there is the slightest doubt about the extra or credit. They do not want to delay the total settlement for "a few bucks." There is a slack in profit margin. (One sub dropped his bill $20.00 immediately. Another said, "Take $25.00 off and if it is too much I'll call you." One sub, on itemizing, found another credit in the patsy's favor for $30.00 when the query was for over $5.00. He proudly pointed out the error, saying he was no gouger.) Thus, if applied, the patsy's edge in power symmetry can save him some money.

Itemizing quickly and clearly, giving the impression of honesty, is the sub's counter strategy to queries and for getting the bill settled normally. He also may follow through with "sweet talking" and explanations of each item for complex jobs as in bidding and blaming crews or estimators for invoice erros. These latter strategies increase the less clear the bill, even when itemized. Subs may also use anti gouging strategies when called to explain a bill by the patsy: they show how inexpensive the job is or how small their profit margin. To wit, the sub is not gouging the patsy. Or a sub may say, "There will be no argument—I'm always fair."

Immediate settlements lead to different modes, because the sub has to break out of the normal mailing thirty-day settlement mode. Mailing takes time and is easily accessible to delays from bookkeeping and management; this is why thirty days is allowed. To pick up the pace of normal processing, subs use various strategies, as we have seen. The bill comes unannounced, the settlement close on its heels. The sub either asks for his money or presents a bill for quick settlement right at the closing of the job, or he returns a day later unannounced, or he calls and asks to come over to give the bill or collect his money. He explains his move "rationally." In many cases the patsy is surprised to be asked to pay so quickly, especially partial payments. He is off balance, so arguments—when he has the money—do not come easily. If possible, to soften the surprise, the sub will make his unannounced request linked with another purpose. He may come back for equipment left there, be in the area to check on what his crew had done, go over extras, or make it part of closure.

Then he asks for a check immediately, makes a date to return for one within a few days, or requests that one will be mailed out within a day. If he does not collect his money quickly, he presses by phone or by returning. He can virtually *stampede* the patsy into quick payment. He may demand closure so hard that it can undermine the patsy's judgment about when closure is appropriate, so he can't judge when it is all right to pay and yields this power to control closure. The sub can complain bitterly about people who didn't pay him, so the patsy feels that he should not be like one of those people. (One sub who wanted to be paid quickly came by to check and see if the bill, sent the day before, had arrived. He was too shy to ask for it immediately, so used this tactic to stimulate an early payment. It did not work since the author was, as most generals, set up to handle bills normally, and broke this routine only under direct request or strategy. He never broke it under interpretation or seeming request.)

Gouging: As noted in an early chapter, the typical imagery that patsies are gouged by subcontractors was not the case in this study. When a patsy has to pay more, it is his cost for a place in the subcontracting systems. It is the only way he can compete for the sub's time against demands from general contractors elsewhere. But a patsy is subject to low range gouging from "extras," (forgotten credits) and "time and material" jobs, which he always has to watch closely so as not to lose his edge in power symmetry. The patsy develops several counter strategies to check and to prevent the ways that the subs gouge. Let us begin with the latter's moves.

The basic move of the sub is to generate some kind of slippage to a higher price from the price as bid. Adding on extras is an obvious way, or forgetting credits. Another strategy is to spend so much time on details as to make the job a "special case," and charge on a time and material basis. The sub itemizes a bill which is an "actual" not "bid" price. He doesn't make this switch unless the actual is higher. The patsy does not pick up the benefits of a low bid. The sub's rationale for breaking out of the bid to a higher price is, "What do you want me to do, lose money? I'm charging you a fair price for a fair job, even if it is over the bid." This, of course, begs the point because the bidding system is based on getting fixed bids. The patsy should benefit if the sub cannot make the bid, thus having to work at a loss for the patsy: it is not up to the patsy to rescue him. This strategy jeopardizes the patsy's trust in the sub, thus hurting referrals. In one case the sub had told the patsy he accepts less than actual costs *after* the patsy has paid.

Also, to try to change a bid to time and material, the sub may use guilt-producing mechanisms on the patsy. A sub may complain to the patsy how surprise costs, union increases in wages, extra time, or increases in material costs are eroding profit—and he must charge more. Or a sub may do favors for the patsy; so when he gouges, the patsy will not complain. His gouging will be averaged out in paying for part of the big picture. Or a sub will "go cheap" on a bid to a point where he can't finish the job until the patsy agrees to pay him for his time and material in order to keep the job going.

These strategies to gain slippage on the bid are designed to swing the balance of power symmetry in favor of the sub from the patsy's power of payment. To maintain his edge the patsy must counter-strategy—if not, he pays the price to settle the closure. The patsy's best strategy is to *argue,* and we saw it used in mode of settling a bill as well. If the patsy is too shy to phone to ask for itemization or to complain, he must pay the high bill. Some subs gouge as soon as they discover a client who doesn't seem capable of arguing. Others gouge everyone knowing that on a probability basis only 25% are capable of arguing.

To avoid argument, the patsy must engage in *pre-bill strategies* and so prevent a gouge. On time and material jobs, he may point out at closure to the sub his *clocking of hours* of the crew's work, as a way of signaling what he will pay for. This is an excellent way to prevent the sub from adding on a few hours to increase the bill, which is so easy and tempting to do.

If the sub queries him in a way that appears to be for ascertaining his ability to pay, the patsy reports how tight he is for funds. Another strategy is to call the sub out of the blue and become *very angry* at the excessive bill that the patsy suspects he will get. This puts the sub on guard about sending too high a bill. To gouge becomes, for the sub, an admission of guilt, since the patsy has defined him as a potential deviant and must be proven wrong. (One confrontation sounded like this, "Don't give me a lot of extras. You gave me a flat price and that's what I want." The result was *no* extras from the roofer for the skylight details.) Asking the sub "If I'll gel a chance to argue before paying" is a strategy that patsies use too, to warn the sub of impending difficulties for excessive charges. (A sub pleaded "always fair" in the face of this; another time "arguments can go either way," silenced the sub.)

Another pre-bill strategy is to request the bill be purveyed, when possible, through another friendly sub, so that the bidding sub must appear reasonable. (Thus the *tree* contractor was asked to send his bills through the excavator, which prevented an excessive gouge. The *tree* man did not want to jeopardize future work from the excavator.)

The last pre-bill strategy is to gain from the crew the sub's true costs for time and material—many crews like to give such information. This makes the patsy ready for argument, if necessary. The patsy may also check material pricing and do a few phone call bids with exact details on simple jobs as a pre-check for a feared, coming gouge, or as grounds for arguing in anticipation of a pending gouge. (For example, it was easy to call around and ask others what a power pole and installation would cost, to discover a gouge by an electrician and to develop steam to argue. The electrician gave him a credit on the coming house wiring after one put-off attempt.)

* * *

Closure on the last job—hardwood floors—brings closure to the total building project, a view which only the patsy, his crew and family can understand and take joy in regarding the months of work it took to get there. Over the past months the patsy has generated a project to success. *The next person on the premises will look at the house as if it had always been there.* Subs, if they see him again, will ask him when the next job, if any, begins so they can bid—not whether the house was ever finished at all. If anyone is to conceive the total project, the patsy must tell them, if *he cares to share his joy.* People will use the house never mindful of who bid and who generalled the dozens of jobs that went into its construction. The fruits of successful generalling are private.

Part III

Concluding Views

IX

Notes on the Expert-Layman Relationship

This study has provided us with a wealth of categories and basic processes pertinent to the expert-layman relationship. To be sure, the topic of this study gives only one set of values to these formulations, but it is easy enough to further extend, generalize and qualify them through comparative analysis with other kinds of relationships.[1] One need only, for example, think of the power symmetry of other expert-layman relationships (such as the well-known ones of lawyer-client, student-teacher and doctor-patient) to discover a fuller range of bases for edge in power symmetry, or to discover conditions under which comparative bidding or generalling are appropriate.

This book has opened up the study of expert-layman phenomena far beyond the standard ones encountered in typical studies by going beyond the conceptions of control, autonomy and helping, to processes of bidding, gaining information, redesigning, inspecting, administering, evaluating, detailing work to temporal pacing, winning trust, obtaining and judging quality and closure, articulating, elsewhereism, generalling and so forth. But even with respect to the standard concerns of autonomy and control, our set of formulations on the patsy and the sub gives us new ideas on what happens when the expert is not at such a great advantage in facing a vulnerable layman, as when doctors, lawyers and accountants face their typical, accepting clients. Also this book points up the often restrictive view of these studies from the standpoint of the limited number of varieties—often preconceived as relevant—but from their limited densification and integration.[2]

Now let us go over some of these new processes and summarize their main hypotheses while generalizing and extending them somewhat through comparative analysis of other professional expert-layman relationships. Obviously I shall only be able to touch on a small portion of the hypotheses presented in this book on expert-layman relationships. And obviously, this chapter could serve as the beginning basis of another book on a formal theory of processes of the expert-layman relationship. Strauss and I accomplished this same transition in going from *Time for Dying* to our book *Status Passage,* where we comparatively analyzed the dying passage with many other status passages. And this author presented a formal theory of *Organizational Careers* by comparing several diverse studies on the subject. The method for this approach is delineated in Chapter 9 of *Status Passage.*[3]

This chapter, then, will be a small, fragmented formal theory of expert-layman relations. It will also illustrate what other hypotheses may be taken from this book and used or developed. Lastly, it will show methodologically how hypotheses developed about facets of the expert-layman relationship here can be broadened, extended and modified to the level of formal theory by comparative analysis with anecdotes and studies of other kinds.[4]

I shall organize this chapter under the three core categories pertaining to experts' relations with laymen: *choosing experts, power symmetry* and *elsewhereism.* The first category has to do with bidding experts, finding them, referrals, and judging whether or not to use experts. Power symmetry develops the traditional notions of autonomy and control on a processual basis which transcends current work. Elsewhereism focuses on the constant competition that laymen face with unseen others in claiming the time and work of an expert. Like the foregoing chapters, this chapter is written, by and large, from the perspective of the involved layman.

Choosing Experts: Bidding

Experts are chosen in several ways that indicate degrees of flexibility of choice.[5] At one end of the continuum is being *forced* through circumstances—like insurance, local monopolies, or public facilities—to accept the expert provided, and in particular, professional experts. In the middle of the continuum, there are experts *referred* both by other laymen—usually clients—and other experts, specialists to specialists. Referrals too can be forced, when only one expert is available for such referrals (e.g., a surgeon), but usually a few are referred from among a much larger aggregate.

Also in the middle of the continuum of the degree of flexibility is *comparative shopping.* When there are many experts to choose from, and many types and sub-types of them, the layman can, over a period of time and many trials, shop comparatively provided he is not hooked by a particular one. Concerning health care, there are many specialist types within medicine for a disease condition, and many experts outside of medicine such as natural health food practitioners of various sorts[6] and practioners of home remedies and myths.[7] As the layman compares results and prices and relationships, he can become quite flexible in choosing whom and how many experts to see.

As we have seen from this study, comparative bidding is a formalized, very flexible method of choosing an expert. It is basically a condensation and codification into a short period of what occurs with comparative shopping, but without the experience of work and the results which come with comparative shopping. As we have seen in Chapter III on detailing, this experience is supplanted by the experts development of trust in the layman that timing and results will occur as he has promised. The expert can go no further in the layman's particular case. The expert must rely on getting referrals from people with satisfied results and through showing his actual past work to the layman. Still, the layman must trust that he will also get the same quality and timing.

A crucial question regarding the use of bidding is, under what conditions will experts allow laymen to bid them? Not all experts appreciate the protection that bidding affords laymen from poor and/or high-priced experts, and the control afforded by the understanding and information gained—with its consequent allowing the layman some ability to inspect and judge quality, and administer cogently his particular work. This, of course, undermines the favorable dominance that professional experts have by virtue of their position and training: they see being a licensed doctor, lawyer or accountant should be protection enough for a client. "Why shouldn't people ask lawyers and doctors for bids?" They can save money and obtain quality without learning the hard way. But such experts do not like being invidiously compared, with a consequent undermining of the power of their positions. They expect to be accepted, *carte blanche,* with trust. They insist on faith. A layman with comparative notions must spend his money for visits while comparatively shopping, or surreptitiously comparatively bidding. Thus the experts do not know and do not give free bids as when there is a bidding system, however informal.

When they are aware of being compared, such experts, who as consultants have given what amounts to another bid, still will not take another's client. "Raiding" is unprofessional. Thus comparative bidding demands an ethical as well as a structural process base. It must be accepted as fair and just, not viewed as undercutting and raiding—as a form of unfair competition among experts. If a doctor does provide bids he is sure to lose his foothold in this referral network,[8] and thus be chastised by losing more clients than he has gained. This makes it hard for a layman who does shop comparatively to bid in order to switch experts who are aware of this. The layman is apparently the first visited expert's client, and a second expert would be fearful of taking over without the first's approval that the switch would be for the layman's good, and on some grounds that neither invidiously compare the two nor breach ethics. Of course, many questions remain for research on what further conditions—besides protection of the layman's interests and undermining of the expert's control—prevent bidding among many kinds of experts. For example, another condition is that bidding gives a degree of organization to an otherwise unorganized aggregate of clientele, which provides a source of lay pressure that professional experts would rather avoid.

Comparative bidding systematizes approaching and obtaining the maximum number of options for what goes on in all but some forced choosing of experts. The search is for protection against exploitation and incompetence, finding trust, understanding the best qualifications, and some control over quality, timing and costs (or some edge in the power symmetry of the relationship).

Whereas professional experts expect imputed trust on the basis of having achieved a licensed position, bidding also provides trust through a build-up of understanding and details through comparisons as well as by licensing. One aspect of this build-up is that the expert can not sustain a *false front* to gain trust, as comparisons allow these laymen easily to see behind it. Being too sweet, double talkish, secretive or haughty is not effective. The next bidded expert probably will show that these styles are put on and are unnecessary to obtain quality. Bidding forces experts to play it straight. The expert wants a job, and then callbacks and referrals; he does not want to be dumped as a comparative phony.

The information gained by the layman during detailing is variously undermining of the trust expected by professional experts who insist on faith, and do not subject themselves to bidding. Information is a kind of evidence that demonstrates competence, and professional experts want this competence assumed without requiring demonstration of knowledge. Information-gaining also undermines the myth of secrecy of the professional's affairs and skills, thus eroding trust in the magic of the expert. Some professionalism is too complex to demonstrate satisfactorily so the professional can only variously impress the clients.

Further, the layman need not trust the expert as much since information gives him several offsetting powers. If the job is simple enough—as much law, accounting and medicine is—he can dispense with the professional and do-it-himself, an action offensive to experts (apart from such crucial exceptions as getting rid of unwanted patients as in some chronic illnesses). Also, information gives the layman the ability to inspect work and temporally administer the expert to a "crowding" degree. Both are forms of control which experts—professional and unprofessional—rarely wish to yield. They think that the layman should follow their schedules and accept work as O.k., not cogently question or control on the basis of either. Even in bidding, experts give as little information as possible on how jobs will be done.

Whereas bidding provides a build-up of trust through comparing, detailing and playing it straight by the expert, in those expert-layman situations in which bidding is taboo and trust insisted upon *a priori,* any trust build-up is negligible until many days after the expert begins work—not, as in bidding, before work. At this point distrust may arise too late or trust be inconsequential; build-up provides only for follow-throughs, callbacks and referrals. For the expert, the positive consequence of the client's not putting trust in his professional skill to a test in choosing him before he has worked is that a doctor, say, can work on acutely ill patients without question, immediately, without being hindered by their subjective judgments. But when patients or clients have time, this delay in establishing evidenced trust until sometime after work begins can be costly to them. Bidding, then, is a sophisticated approach to resolving the dangers of haphazard, fortuitous choosing or long-term shopping. The lack of protection brought on by premature trust not only allows exploitation based on trust—the least danger—but allows being provided less-than-competent quality and skill, or provided with the wrong expertise, or delays in finding the correct specialists. Bidding limits these risks.

Power Symmetry

Comparative bidding provides the initial basis for a near-equal balance in the power symmetry of the expert-layman relationship. Using comparative bidding, the layman develops good judgments on what to expect, to demand and pay for quality, timing and cost. This amounts to control over the expert. In contrast, the professional-client relationship is usually quite asymmetrical in favor of the professional, since it takes time to develop these judgments to a cogent degree. And there is seldom the time necessary until long into the course of a single, non-bidded, expert-lay relationship, even if such judgment eventually develops. Indeed, many patients are still ignorant of how to judge time, quality and cost after years with one doctor—they lack a comparative perspective. In spite of having the time, they have no process, such as comparative bidding, by which they can systematically develop judgment.

The problem of power symmetry is complex. One cannot assume that an expert deserves a complete edge in power because of his position. (He may be an "inexpert expert" facing a knowledgeable layman, and his domination may not be just.) Conditions which are acute and serious, requiring instant expertise, demand an asymmetrical power relationship in favor of the expert: he is given *carte blanche* to act, as in emergencies. But when matters are not serious—though acute—or are long-range or chronic, then questions arise as to the justice in giving an expert such power asymmetry without proof of merit.[9]

But more to the point: to take over the job of expert or to control an expert, *a layman need not be an expert in general but only in his specific case.* There *is* enough expertise for him (but *not,* of course, enough for the expert). Indeed, on this detailed level, when the layman may know more than the expert, his judgments should, reasonably, prevail. This is starting to occur in cases of chronic diseases and in civil rights cases where intelligent or busy professionals allow or even advocate their clients' views.

It is from the asymmetrical professional-expert relationship that much of the "know-it-all/know-nothing" imagery emerges, with its consequent instant faith and yielding of dominance. *The mirror image of being a professional expert is being an obedient layman to other experts.* This is so inbred in professional experts that often they make the best patsies to experts in other fields! Thus a doctor turns himself over to a lawyer with complete, unquestioned trust on the basis of a mere referral. Knowing the patsy role through his own expertise, it is easy for him to do it when facing other experts. On the other hand, the bidded relationship teaches us how experts may become laymen to other experts so as to yield enough control while limiting domination although they are out of their own bailiwick. It gives to the relationship power symmetry while trimming or doing away with unquestioned trust and its contingent unprotectiveness and paranoia for the expert-turned-layman.

Imagery of what work is to be done by an expert is vital to its accomplishment. In an asymmetrical relationship based on instant trust, this imagery is not problematical during the beginning of the work. The layman makes no pretense at imagery and, with the barest details of his desire, leaves his fate in the expert's hands. But when power symmetry is a slightly shifting edge in balance, the respective image of a client and expert becomes problematic and possibly conflicting. This is why detailing is extensive in bidding—so that layman and expert both have a clear view of what is to come, and both agree. Obviously, professional experts prefer as little detailing as possible—even complete nondisclosure—so the layman's imagery will not conflict with and erode their power advantage. Such details as are disclosed are at minimum, and legitimated by them alone.[10] This minimizes their loss of control in giving details, hence understanding, to the layman.

Imagery in the symmetrical expert-layman relationship, then, may easily lead to differential judgements *that matter* for the future course of the work—unlike the asymmetrical relationship. To keep the work going, as we have seen, the resolving of these differentials must be paced, and carefully, by each party. Many strategies (delineated throughout the book) resolve those differences. These are the strategies of reduction of anger or concern, delays, discounting, listening and then doing what one considers best; also redesigning, intense detailing by the expert and by the layman, generalling strategies of pacing, generalling of orders, inspection and evaluations to the point where experts will not be stopped or quit (but do not proceed unfettered), playing patsy, information gaining, playing wise, pressuring, articulating, correctly paying, dismissing, controlling before being controlled, and negotiating, and so forth. These strategies do not come easy for many laymen who are accustomed to asymmetry in client-expert relationships.

But these same strategies, while resolving, *may also be used to generate* differential judgments and desires of imagery, which process can turn an asymmetrical relationship into a balancing symmetrical one if the layman has the temperament to try. For example, if a patient just gives orders out-of-hand to a doctor or lawyer, this would probably result in the expert's using the strategies of discounting and reducing with no allowable affect on asymmetry. But if the layman paces his orders to vulnerable junctures, such as at the end of detailing the case, or when a preliminary solution begins to be less than successful, or directs his orders at a para-expert in service of the expert; then he can start balancing the symmetry and developing control.

The layman cannot be shy nor waiver in telling a person who knows more than he what he ought to do. The layman must draw on his power to know his own specific case best—both technically and when applicable, with taste—and his power to reward, and his power to go elsewhere.

Part of pacing his orders is learning to leave an expert alone when he is working well and has sufficient imagery. To say something may stop the expert, and hinder the achievement of the layman's goal. The goal is *not* to dominate, but to achieve a goal with some controls in the layman's favor to secure what he wants. By definition, the expert is working for the layman's desires (though one can sometimes wonder about this in some asymmetrical relationships).

In our books on dying,[11] we have shown how the hospital staff manages patients for many purposes other than patients' welfare, e.g., nurses require an orderly ward too! The same process can apply in other client-expert relationships concerning administering and timing. When a layman wishes to right an asymmetrical relationship in order to gain some control, to be effective he must carefully gain enough information and pace his actions and words with the course of the expert's work. Inspections should be based on near-completions, or for corrections, or on the verge of errors—not before the expert has produced a viewable result, otherwise the layman's action will be premature and discounted. Administering should be attempted when temporal pacing is off, as for delays or for unhappy consequences for the layman elsewhere. Again the focus is on the goal—not on dominance—or the expert will not yield.

Well-paced interventions by laymen proportionalize the expert's imagery on the work, as the balance of power shifts toward symmetry. Give and take begins, pre-empting of control ceases; differential judgments emerge and are resolved. If enough laymen were to try to balance asymmetrical expert-lay relationships, they could change professional training and the professional practices. This is happening in some measure, as minorities, imbued with civil rights, demand better medical care. It happens as malpractice suits increase and as an individual's rights are more guaranteed, as in the commitment of possibly insane people. Taking the asymmetry out of many expert-layman relationships is beginning to be frequent among the rebeling young in the realms of education, occupations, economics, health and politics. However, the balancing out of power asymmetrical relationships still does not provide the layman with the capability that we have seen in this study of being able legitimately to general many experts on an articulating basis. The structure of professional relationships is not often suited to the generalling of many lawyers or doctors. What generalling and articulating by a layman does accomplish is not manifest to those experts who are unaccustomed to fitting their own schedules to the bigger picture of a layman. The layman must do the best he can while keeping his own projects fairly out of issue. The layman in the professional relationship must depend on strategies, not on social structure, to gain and keep some control.

Elsewhereism

This study offers a new variable of major importance in the course of expert-layman relationships and their outcomes: relative to one client, competitive demands from elsewhere upon the expert—or elsewhereism.[12] Stories are legion among laymen about delays from waiting for or missing their busy expert—and how fateful that is for them. I am reminded of a Los Angeles Times story about the woman who called her doctor, who was too busy elsewhere to see her. As her need was acute, a substitute doctor was sent by the answering service: he then hooked her on heroin and kept her hooked for many years! It is my contention that sociologists cannot talk of experts' servicing laymen without talking about how the service is articulated with demands from elsewhere, and how these demands affect quality and closure of servicing work. Busy doctors and lawyers often spread too thin are a problem to themselves and to clients regarding their timing and quality of service. Getting needed appointments by patients is a classic example of the problem. For example, one research study shows how people needing immediate psychiatric care are made to wait for two weeks by a busy clinic—over which time the problem disappears or changes drastically in the patient's view.[13]

Also it is my contention that *how* a layman competes for an expert with laymen elsewhere—usually unknown, given the power symmetry of the relationship—is important for understanding the course of an expert-layman relationship. For example, in our article on the social loss of dying patients,[14] we state that social loss may be a determinant of the amount and speed of medical and nursing attention. This proposition would have been more to the point had we followed through—on the hypothesis that people with low social value lack staff attention, with the idea that it is because they compete poorly with other people who are of a higher social value, of with experts' other affairs of more concern. This idea was surely documented in the data.

Elsewhereism primarily affects the pacing and articulation of the expert's work. As we have seen, this timing then can have quality and cost consequences. Either the expert has no time or he is delayed, cuts short his work, or finishes too fast because of demands from elsewhere—and the layman is in poor competition with these demands. If the demands are minimal or well articulated, however, the layman experiences no effect from elsewhereism. Inasmuch as most of these demands are not minimal and articulate, the layman is bound to feel elsewhereism, at least through short delays and often long ones. It is difficult for the expert not to pass on to a client the effects of elsewhereism: he can only cover so much with other experts or crews. Thus, when doctors have back-up colleagues, the colleagues too may be hard pressed by elsewhereism to back up the expert with his demanding client.

To the degree that elsewhereism is passed on, the layman must compete with other clients, with little or no knowledge of who are his competitors or what are the criteria of competition. He only finds out the barest information when the infrequent expert defends, explains or excuses his delay, and tells what had been going on elsewhere. The criteria of competition may vary from realistic to very difficult or impossible to meet. A mildly ill patient cannot win out against an emergency—he must wait. Patsies can outdo generals with prompt cash payments. The more an expert talks of his clientele, of course, the better astute laymen can learn how to compete for his time. An expert seldom talks this much, especially since it is information confidential to the professional relationship. Perhaps the best way to compete against high odds is the threat to go to another expert, to become an emergency, or to wait for a slack season. It is hard to get an accountant to do other work during "tax time," but afterwards he will. Many professional experts have slack periods.

The layman develops competitive power symmetry when, as a general, he must articulate several experts on the job—a structural property of the patsy-sub relationship—by delaying some experts and affecting others. But in professional expert-lay relationships and other kinds of servicing, where uneven balanced power obtains, the power gained from articulating is not as much since the layman does not general and articulate the work of several, say, lawyers on a legitimate structural basis.

A major property of elsewhereism, then, is the degree to which a layman is aware of with whom he is in competition. The practice of most professional experts is to say nothing respecting identity of their other clients, so clients must compete as best they can in the dark. As I have said, experts often only divulge properties of competing clients when excusing their tardiness. Most client-sets are an unknown aggregate to each other. But several structural conditions may obtain when knowledge of other clients becomes possible, Lay referral systems give this kind of knowledge on other clients as well as information on the expert, as do professional referrals although to a lesser degree. Information is given on the type and class of clients, not on particular ones, as in the lay system. Another condition is when many clients are grouped together in full view of each other such as in waiting rooms or in wards. The client can make judgments as to competitive ability in making appointments and warding off future delays. Unlike waiting rooms which just present an aggregate of types, hospital wards present a type of client, so competition criteria are clearer. If a client does not anticipate doing well, we can change wards; a medical or surgical emergency patient does not do well on a psychiatric ward. The less a layman knows how to compete, the more the expert can maintain his edge in power symmetry over him on when to see him and for how long.

When a layman is not getting the attention promised and needed, he can accurately assume he is not competing well with clients elsewhere. There are several conditions ascertainable by the layman that make for stiff competition which, with professional experts, usually is only of hourly or daily duration. The "best" experts in any field have several "prominent" clients, "good business" clients, or "deserving" clients who can easily crop up and take priority over an average client whatever his need and protest. Some clients make a practice of working for the position of a high priority client in order to obtain instant attention: they do not wish to worry about elsewhereism. For example, a corporation or businessman will flood a lawyer with business so he would not dare not to respond with instant attention to his slightest call. In consequence, a smaller client with a one-shot problem may find himself repeatedly bounced from the focus of attention. He, then, must pay for his place in the expert's clientele (and the larger system) with time and/or money. He must judge whether this "best man in town" expert is worth his time and money.

Another source of stiff competition knowable by the layman is the expert who is too busy by allowing himself to be spread too thin, whether or not he is among the best experts. He presents the layman with a constant rat-race of elsewhereism. No matter how well he articulates his practice, the large number of wanting clients are always interrupting and demanding. The phone is constantly ringing, the temporal demands are spotty and of short duration, but derailing for the layman who is seeking his expert's service. Even for crucial help, he can be made to wait a moment by the expert.

Other temporal conditions are the peak load hours and holidays which generate elsewhereism problems for any layman, and even for those experts who articulate client-sets well. Laymen generally avoid these times, unless they too become deserving clients—that is, feel that their sudden problem will compete well—or are forced to call an expert. Another temporal condition is the geographical swing of the expert and layman on the *fringes* do not compete well with those near his geographical center. An expert can see more people in the center: temporally it is costly to go to the fringe to see one layman.

Another condition that generates baffling, stiff competition for the layman is the "squeezability" of elsewhereism. Whether it be a doctor, lawyer, accountant or sub, the question arises just how well, given the type of expertise and practice, the expert can squeeze in another client for a quick phone call or job while attending to a client's needs. High squeezability may allow an expert to take care of many other clients while the primary client does not know the difference: for example, in the well-articulated office, where the doctor goes from examination room to examination room while patients are being prepared. Low squeezability leads to constant interruptions and delays if the expert tries little jobs or calls in between and spreads himself too thin. The principal consequences of elsewhereism, as we have seen in Chapters IV and V, are delaying the layman and other experts. And, of course, delays also have their consequences.

In sum, besides what knowledge the layman may glean from the expert and the viewing of other clients which enables him to expect elsewhereism occurrences, he may also expect them on the basis of several conditions attending the practice of expertise. The expectation of elsewhereism is important because it causes delays in work or causes poor quality. These consequences in themselves have chain affects on future work, total projects, and other people (elsewhereism) whose fate is linked with the work. Delays of doctors are continually altering patients' schedules and subsequent experts do a good business in correcting previous experts' quality, some of which is caused by rushing.

As we have seen, in experts' work, delays may be caused by elsewhereism at any point: start, during and closure. The occurrence of elsewhereism is usually unpredicted exactly, but in general is expectable. The expert will warn if possible—the secretary calls—but in professional expertise he is often caught and cannot get away or even get to a phone. Thus the layman can expect surprises. Some professional experts are so used to such delays that their clients are never informed but simply left to understand on their own as the patients or surgeons or gynecologists. Experts, then, vary according to the mechanisms that they set up to inform laymen of elsewhereism demands which formally excuse delays. The mechanisms are associated with reduction and hooking strategies as discussed in Chapter V.

In response, the layman responds with some combination of understanding and anger. While anger gains some edge in power symmetry for the layman, in asymmetrical relations with professional expertise, anger usually does little good and the layman is left to understand or go elsewhere. If he cannot withdraw, he must rest and wait despite his anger. Withdrawal is harder when the expert and layman are well into the course of work. This is another reason why laymen lose their edge in power symmetry during the middle of work. It is at the beginning and at closure that they develop most of their power in symmetrical as well as asymmetrical relations.

We have seen that delays when expectable and of short duration generate a zone of tolerance and normality and do not occasion particular strategies beyond an attempt at understanding. But as the zone is breached, the layman will start ensuring an end of the delay by using strategies like reminding or verifying of time, or seeking a reaffirmed commitment. (For example, patients check with the secretary on whether they came at the right time.) If this does not eliminate delays, then "pushing" strategies come into play—the client applies direct demands and pressures through stressing imperative needs, making threats to withdraw, confronting over the breach, or making demands to know why the delay? These strategies are familiar to any professional-client relationship. Lastly, anger may be used in full force irrespective of the consequences for the layman. The anger to be effective must be controlled and paced to gain an edge in power symmetry, otherwise it has no affect on the delay and may even occasion more; as when a patient is considered a "nut" and is custodialized by para-experts while waiting, or when a defendant is denied audience with counsel until he calms down.

The handling of delays become more complex when a layman must work with many experts (Chapter VI) not an usual condition in professional relationships of chronically ill patients. Here the problem is articulating a life-style elsewhere to jibe with the ever expert's schedule, which is seldom designed to consider the layman's elsewhereism. One can see that the problems of elsewhereism become quite complex when both layman and expert are articulating demands from elsewhere in order to work together. The typical first strategy is that of appointment-making, which the layman articulates to the expert's articulation because of initial asymmetry in power: the layman needs, the expert is busy.

I have barely touched the problems of temporal impact of elsewhereism brought out in this book and in considering comparatively the professional-client and other servicing relationships. I have merely opened the way to the problem but before closing this section I wish to touch briefly on the duality impact of elsewhereism. Experts set standard time periods for providing adequate quality, and these times are articulated with promised work for people elsewhere. The professional schedules of doctors and lawyers mirror this, and then the schedule cannot so mirror, these professional charges for time and let the job run. Lawyers' fee for service charges can become huge for open-ended care. Some clients elsewhere may demand excess time in order to get quality—emergency or surgery or lawsuits or accounting snags. To prevent disarticulation of the whole client—set, the expert may take time from jobs that can be "safely" rushed. The expert tries to keep up by doing less well on jobs with no exact quality requirements. Hence quality suffers and a layman is at least partly abandoned. Subs have more exact quality standards that are subject to the inspection of laymen and officials almost instantly. Professional quality is less clear and takes more time to emerge as being poor and perhaps worth correcting. Layman inspection of professional work is frowned upon, as we have seen. Even formal inspections by officials and other experts is discouraged, although some hospitals are developing genuine committees of peers to inspect certain treatments and procedures of surgical operations. Thus, criteria and inspection of quality vary widely in expert-layman relations; hence the effects on them of elsewhereism are varyingly difficult to ascertain. And the further consequence of gaining edge of power by laymen who have found poor quality is correspondingly harder in professional cases of low clarity.

Quality is intimately linked with closure: the two temporally intersect. To be finished on time is to have reached the desired degree of quality. Closure raises questions of guarantee of quality. Whereas subs must guarantee by statute, professionals have no such problem—save layman-instigated lawsuits. Once finished they are seldom held accountable for errors or corrections, sometimes they charge again if they do them. Thus, in professional relationships allowing closure is more hazardous for laymen. The increase in professional malpractice suits only indicates the problem, it does not resolve it. Professional responsibility and controls do not solve it either.[15] Formal inspections and legal guarantees are solutions, to be used or avoided depending on the expertise. They are used for non-professional servicing in areas. Professionals have avoided them, so elsewhereism can have more of a drawing away effect on quality. Perhaps the only control that laymen have is through withdrawal and lay-referral systems.[16] One source of continued asymmetrical power in professional-clients relationships is this lack of legitimized inspection of quality. That which occurs is *ad hoc,* specific, informal or consists of lawsuits which drag on and on.

In conclusion, this monograph indicates that in the explanation of behaviors occurring in expert-layman relationships—professional, non-professional, and many servicing relationships—three core explanatory variables are the processes of *choosing,* the balancing of *power symmetry* between expert-layman and the prevalence and force of *elsewhereism.* Using these complex variables, the sociologist can account for much if not most of the variation in expert-layman behaviors. This monograph also indicates that to use one without the other (such as the intense focus on autonomy reflected in the literature) is to miss the basic explanations of and belief in the complexity of what one is observing. The question is not just the autonomy of the professional, but this variable (which pertains to a dimension of power symmetry) in conjunction with choosing and elsewhereism—both of which affect the degree of autonomy as an expert's work progresses.

Notes

1. See Barney G. Glaser and Anselm L. Strauss, *Discovery of Grounded Theory,* (Chicago, Aldine Publishing Co., 1967) Part I and Barney G. Glaser and Anselm L. Strauss, *Status Passage,* (Chicago, Aldine Publishing, 1971, Chapter 9 and Barney G. Glaser, *Organizational Career: A Sourcebook for Theory,* (Chicago, Aldine Publishing Co., 1968) Introduction.
2. Barney G. Glaser, *Review:* of Eliot Freidson's *Profession of Medicine: A Study of Sociology of Applied Knowledge* (New York; Dodd Mead & Co., 1970), in *American Journal of Sociology,* (Sept. 1971) pp. 361–2.
3. *Ibid.* See also Anselm Strauss. *The Contexts of Social Mobility* (Chicago: Aldine Publishing Co. 1971).

4. On how fragmented theory can be broadened, see Anselm Strauss, "Discovery New Theory from Previous Theory" in Tomatsu Shibutani (ed.) *Human Nature and Collective Behavior,* (Englewood Cliffs, Prentice Hall, 1971).
5. On choosing professionals see Eliot Freidson. "The Impurity of Professional Authority" in Howard S. Becker et al, *Institutions and the Person: Papers Presented to Everett C. Hughes* (Chicago, Aldine Publishing Co., 1968) pp. 27–31.
6. Richard R. Hanson, and Julius A. Roth, *In Quest of Optimal Health: The Natural Health Movement in the United States,* unpublished manuscript.
7. David - Hayes - Baptista, "Judging the Medical Professions: Primary Concern by Chicago", unpublished paper, University of California, San Francisco.
8. On referral networks in medicine see Eliot Freidson. "The Profession of Medicine: A Sociological Analysis" (New York, Dodd, Mead, 1969), Part III.
9. The moral and consequential justification for such unfettered asymmetrical power in favor of the professional expert is dealt with at length in Freidson, *op. cit.*
10. See Barney G. Glaser and Anselm L. Strauss, *Awareness of Dying,* (Chicago, Aldine Publishing Co. 1965) Part II.
11. *Op. cit.* and *Time for Dying,* (Chicago, Aldine Publishing Co., 1967) and *Anguish: A Case History of a Dying Trajectory* (Mill Valley, The Sociology Press, 1970).
12. Elsewhereism effects the shape of the status-passage of an expert's client. See Barney G. Glaser and Anselm L. Strauss, *Status-Passage* (Chicago, Aldine Publishing Co. 1971) Chapter 3.
13. A study by Evelyn Peterson of making appointments for psychiatric help at a clinic by working class people presented in seminar at University of California Medical Center, San Francisco.
14. Barney G. Glaser and Anselm L. Strauss, "The Social Loss of Dying Patients," *American Journal of Nursing.*
15. See Eliot Freidson, "Processes of Control in a Company of Equals, *Social Problems.* XI (1963) pp. 119-131 and Joanne Scherr, "Reciprocal Dependence, Control and Peer Interaction: A Case Study of a Group Medical Practice," unpublished paper. University of California Medical Center.
16. On this see, Eliot Freidson, *Patients' View of Medical Practice* (New York: Russell Sage Foundation, 1961) Chapter 10.

www.ingramcontent.com/pod-product-compliance
Lightning Source LLC
LaVergne TN
LVHW011131300926
849160LV00008B/603

* 9 7 8 1 4 1 2 8 5 7 1 7 8 *